HOW TO RAISE
SHEEP

EVERYTHING YOU NEED TO KNOW

BREED GUIDE & SELECTION
PROPER CARE & HEALTHY FEEDING
BUILDING FACILITIES & FENCING
SHOWING ADVICE

Philip Hasheider

Voyageur Press

DEDICATION

This book is dedicated to my wife, Mary, whose suggestions and encouragement
through the years have made my writings much better
than they would otherwise be.

First published in 2009 by Voyageur Press, an imprint of
MBI Publishing Company, 400 First Avenue North,
Suite 300, Minneapolis, MN 55401 USA

The information in this book is true and complete to the best
of our knowledge. All recommendations are made without
any guarantee on the part of the author or Publisher, who
also disclaims any liability incurred in connection with the
use of this data or specific details.

FFA® and the FFA® emblem are registered trademarks of the
National FFA Organization and are used under license.

Use of the FFA mark and name does not represent an
endorsement by the FFA of this product or an endorsement
by FFA of the manufacturer or provider of this product.

The contents of this book were reviewed and approved by
Dr. Clint Rusk, Associate Professor in the Youth
Development and Ag Education Department at Purdue
University, in accordance with industry standards.

We recognize, further, that some words, model names, and
designations mentioned herein are the property of
the trademark holder. We use them for identification
purposes only.

Voyageur Press titles are also available at discounts in bulk
quantity for industrial or sales-promotional use. For details
write to Special Sales Manager at MBI Publishing Company,
400 First Avenue North, Suite 300, Minneapolis,
MN 55401 USA.

To find out more about our books, visit us online at www.
voyageurpress.com.

Library of Congress Cataloging-in-Publication Data

Hasheider, Philip, 1951-
How to raise sheep / Philip Hasheider.—1st ed.
      p. cm.
Includes index.
ISBN 978-0-7603-3481-2 (sb : alk. paper)
1. Sheep. I. Title.
SF375.H344 2009
636.3—dc22

                                    2008038562

Editor: Amy Glaser
Designer: Chris Fayers

Printed in Singapore

CONTENTS

ABOUT THE AUTHOR

Philip Hasheider grew up on a diversified farm in southcentral Wisconsin where he was involved in 4-H and FFA. In recent years he has combined his interests in agriculture and history to write nine books, including *How to Raise Cattle* and *How to Raise Pigs*.

Philip was the recipient of the 2005 Book of Merit award presented by the Wisconsin Historical Society and Wisconsin State Genealogical Society. He has written numerous articles for national and international dairy breed publications. His diverse work has appeared in the *Wisconsin Academy of Review*, *The Capital Times*, *Wisconsin State Journal*, *Sickle & Sheaf*, *Old Sauk Trails*, *Sauk Prairie Area Historical Society Newsletter*, *Sauk Prairie Eagle*, and *Holstein World*. He was the writer for the *Wisconsin Local Food Marketing Guide* for the Wisconsin Department of Agriculture, Trade, and Consumer Protection.

He currently farms near Sauk City, Wisconsin, with his wife and two children.

ACKNOWLEDGMENTS

Our son, Marcus, helped with much of the photography that appears in this book. It is his third assistance in this series. Our daughter, Julia, willingly participated in helping with the cover photo. Both children are active 4-H and FFA members, much to the delight and satisfaction of their parents.

Several people deserve my sincere thanks for their assistance to my requests for help. Ray and Alice Antoniewicz, Kim and Jeanne Radel, Brenda and Robert Tobey, Cynthia Allen, Dr. Greg and Vicki Brickner, and Tom and Laurel Kieffer unselfishly offered their flocks and farms as photo subjects. Without their help, this book would not be as complete.

I also wish to thank Dr. Dave Thomas, professor of sheep production, University of Wisconsin-Madison, and sheep specialist for the Wisconsin Cooperative Extension Service, for his valuable assistance which he freely gave while under other time commitments.

Todd Taylor, shepherd at the University of Wisconsin's Arlington Research Station, is actively involved in sheep instruction, research, and extension activities. I wish to publicly thank him for spending time with me at his station and flock. Todd shared his recordkeeping model and portions of it appear in this book.

There were many others whose assistance was essential to the completion of this book including Sue Carey, Sherry Carlson, Carol Elkins, Anne and Richard Gentry, Nancy Hann, James Harper, R. J. Howard, Gary Jennings, Don and Janice Kirts, Brian and Jennifer Larson, Rinda Maddox, James Morgan, Shannon Phifer, Steve Pinnow, Mickey Ramirez, Nikole Riesland-Haumont, Jan and Coby Schilder, Bill and Liz South, and Di Waibel.

Jerry and Ruth Apps have been mentors and friends for many years and have always taken an interest in my writing. Their encouragement was the catalyst to embark upon this third book of the series. To simply say thank you seems wholly insufficient.

A special thank you to my editor, Amy Glaser, who provided me a third opportunity to work with her and Voyageur Press; both proving to be excellent partners.

AN INTRODUCTION TO RAISING SHEEP

Sheep have been closely intertwined with human survival and prosperity for thousands of years. Sheep were one of the earliest domesticated animals. Their relative lack of aggression, manageable body size, social nature, and high reproductive rates made them particularly adaptable to human civilizations. From the earliest recorded mention of today's sheep ancestors to the present, sheep have provided humnas with meat and milk for nourishment, wool and skins for clothing and warmth, and a form of value in terms of tribal, communal, or personal wealth.

Sheep have had connections with human civilizations in cultural, religious symbolism and ritual, particularly in the Abrahamic faiths and Christianity. In the Western zodiac, the ram is the first sign and is known as Aries. In ancient China, the sheep (or Ram) was seen as a lucky sign and, as the eighth sign of the Chinese zodiac, represented continuing prosperity. The Chinese believed that good fortune smiles on anyone born in the year of the sheep; they are said to possess a peaceful nature and a kind heart, and are gentle and compassionate, almost to a fault.

As human endeavor has evolved, so have sheep. As man traveled in a nomadic life, sheep became spread across the globe. Today it is difficult to find regions where sheep do not or cannot exist. Sheep adapted to new climate conditions where they were introduced and learned to survive on the local available vegetation.

SHEEP IN HISTORY

Sheep are thought to have originated in the high plateaus and mountains of Central Asia and were domesticated about 10,000 years ago. Archaeological excavations of Iranian sites suggest that woolly sheep were being selected as far back as 6000 B.C. During the Bronze Age (about 3000 to 1600 B.C.), sheep with all the major

Sheep are among the earliest domesticated animals and have provided meat, milk, wool, and skins that have aided human survival. Sheep have a unique history as a form of wealth and for human sustenance.
Sue Carey

features of modern breeds were widespread throughout Western Asia. The Greek and Roman civilizations relied on sheep as primary livestock. The Romans were important agents in spreading sheep production throughout the continent.

It was once thought that today's sheep were descendants of two different kinds of wild sheep: the urial that lived in Southern Asia, and the mouflon, a wild sheep living in Europe. This has been called into question more recently because the urial, argali, and snow sheep have a different number of chromosomes than other sheep species, which makes a direct relationship implausible. Some genetic studies have shown no evidence of urial ancestry. This leaves some uncertainty as to the exact line of descent between domestic sheep and their wild ancestors.

Domestic sheep have slowly and carefully changed from their wild ancestors. Wild sheep were originally tamed for their hides, for milk, and to carry loads. It was not until about 3500 B.C. that they became important for their fleece. What was originally a coarse coat of wool in wild sheep was slowly replaced through breeding by the soft, downy undercoat, which gradually became wool under domestication. Only in the last two hundred years have sheep breeders developed an animal primarily used for meat.

As the sheep population expanded into the Middle Eastern areas, Africa, and Europe, their wool became one of the first products to be valued in international trade. By 1000 A.D., England and Spain were recognized as the major centers of sheep production in the Western world and their wool trade created great wealth for their countries. Both countries had a significant influence in the development and spread of the sheep industry.

In North America, the first domestic sheep arrived with Christopher Columbus during his second voyage in 1493. Other flocks arrived with later explorers who dispersed them throughout the southwestern parts of the present-day United States.

Sheep were imported to the early colonies on the eastern seaboard and grew to a major industry largely due to the political unrest and civil wars in Britain that disrupted maritime trade. The colonists found the demand for wool was easier to satisfy with their own production and soon developed major areas of sheep production.

The British government attempted to stifle the threats to its wool trade by banning exports of sheep to the Americas. This was one of the many restrictive trade measures that precipitated the American Revolution.

As the westward movement took settlers to new areas, sheep went along. The competition between

Sheep can be found in almost every corner of the world because they are very adaptable to different climates, management styles, and vegetation. There are a wide variety of breeds to choose from and learning which one fits your situation best can become a family project.
Jan and Coby Schilder

sheep and cattle ranches grew and eventually erupted into range wars that sporadically occurred between the early 1870s and 1900. The stakes included more than water rights, and the disputes were symbolic of the vast differences that existed between sheep and cattle producers.

SHEEP NUMBERS AND CLASSES

It's estimated there are well over one billion sheep being raised worldwide. The ten leading countries in sheep production are China, Australia, India, Iran, New Zealand, the United Kingdom, Sudan, Turkey, South Africa, and Pakistan.

In the early 1800s, the U.S. sheep population stood at approximately seven million. Sheep had become a fixture on most farms and even populated the White House lawn during World War I. Sheep numbers peaked at about forty-five million in 1945.

The 2006 United States Department of Agriculture census recorded 69,090 sheep farms or ranches. Small producers (those owning less than one hundred head of sheep) comprise the majority of sheep enterprises but own about 17 percent of the total number. Large ranches

own approximately 80 percent of the nation's flock and are primarily located in the western states.

According to the January 1, 2008, agriculture census, there are 6.06 million head of sheep and lambs in the United States, 2 percent lower than the previous year. Breeding inventory totaled 4.51 million sheep, while market sheep and lambs stood at 1.55 million.

The western U.S. range lands still support the production of large numbers of sheep flocks. The ten major sheep-producing states in the 2006 USDA National Agriculture Statistical Service include:

State	Number of Head
Texas	1,070,000
California	610,000
Wyoming	460,000
Colorado	400,000
South Dakota	380,000
Utah	295,000
Montana	290,000
Idaho	260,000
Iowa	235,000
Oregon	215,000

Sheep are hardy grazers and can make good use of grasses and pastures that are difficult to mechanically harvest. Fields with stones, brush, or trees that make them unworkable are still very accessible to sheep.
Shannon Phifer

Gamboling sheep in a pasture is a pleasurable sight and can enhance your decision to move to the country and embark on raising sheep. Not all sheep act as high flyers but their actions suggest a healthy physique and a sense of humor.
Cynthia Allen

In recent years, many small-scale sheep-raising programs have emerged either as a result of or in conjunction with niche markets involving artisan wool, cheese, and meat products. Sheep have become popular on farms using sustainable practices and with those people desiring a country lifestyle with a hobby or retirement activity. Whether or not you fit either of these categories, sheep farming can be a rewarding and satisfying experience. The prolific nature of sheep is such that with good management, adequate nutrition, and attention to basic animal husbandry, you may see your flock expand greatly in a very short time to provide many options for your farm and program.

Farming can open new possibilities for you and your family. It can be emotionally rewarding to have a close connection with the land. However, moving to a rural area also brings responsibilities, such as land stewardship and helping maintain the rural character for future generations.

WHY RAISE SHEEP?

The ways in which sheep are raised have changed little over the years. Because they are a ruminant, sheep can survive in any climate where grass or pasture is available. Feedlot sheep are a possibility, but open areas for grazing provide the healthiest atmosphere for them to thrive.

There are advantages to raising sheep over other farm livestock, and your situation may be favorable to include them in your farming enterprise. Raising sheep can be a pathway for a young person's entry into 4-H

or FFA livestock programs because they require little investment and can produce a saleable animal in three to five months' time. Raising sheep can provide a source of food for the family, as well as a sense of satisfaction that comes with working with an intelligent animal.

Sheep do not require extensive housing or equipment to be raised in a humane and safe manner. With a little guidance and sufficient knowledge, any fledgling sheep grower can successfully raise sheep from birth to market. Sheep are among the easiest animals to raise if

You do not need a large acreage to farm or raise sheep. However, having access to pasture areas will greatly enhance your ability to raise sheep more economically than if fed in small lots. As ruminants, sheep are natural grazers of grass and brush areas that can help control unwanted weeds and other vegetation.

you're new to the countryside and wish to embark on a small livestock-rearing enterprise. Sheep are ideally suited for small-scale farms because of their manageable size, gentle nature, innate intelligence, ability to consume feedstuffs efficiently, and low initial investment.

There are many sound reasons for choosing to raise sheep instead of other livestock. Reasons may range from a homegrown food source for your family to a pleasurable pastime and the aesthetic value of having a pastoral, gentle animal walking on your land. The sight of sheep quietly grazing on a hillside or in a pasture can be a soothing scene for many people.

Raising sheep has several advantages when compared to raising other species of livestock, and most of these advantages relate to the investment of money and time. By the time a lamb is several weeks old, it can be fairly self-sufficient, although by its nature it will still seek its mother. A lamb can feed and water itself with little assistance and be off to market or harvested in less than six months.

Lambs are cute and playful when they are very young but they will quickly grow into more substantial animals. The appearance of these Cotswold lambs will change significantly over the next year. *Bill and Liz South*

Although there are many sheep breeds to choose from in the United States, you may be able to secure new genetics by importing semen from other countries where importing live animals to secure a breed is difficult or impossible. This Gulmoget originated from semen imported from the United Kingdom and is one of thirty recognized varieties of markings in the Shetland breed.

In contrast, a beef calf can require twelve to fifteen months of care before it reaches market weight. A dairy calf is still an infant for many months and needs special feeds, such as milk or milk replacements, to ensure sufficient growth until its rumen develops to where it can eat solid feed. A dairy cow requires considerably more care and feed, plus facilities and equipment for a milking system.

With more than fifty breeds and types of sheep available in the United States, you can easily find a breed with characteristics that best fit your farm and aesthetic tastes.

SHEEP FOR PROFIT AND PLEASURE

During the 1960s, a shift in specialization on farms occurred and farmers moved away from raising many species of livestock, such as chickens, swine, beef, dairy, and sheep, to only one or two kinds of animals.

This trend paralleled the rest of agriculture that was evolving into what became known as agribusiness. Farms

Jacob sheep are one of the oldest sheep breeds associated with human history. You may consider raising a heritage breed on your farm to help maintain or increase the breed's population. *Mickey Ramierz*

became larger and farmers devoted more time and expertise to fewer species of livestock. Similarly, as price fluctuations and profit margins altered different livestock markets, specialization in one or two species became the norm. There was a noticeable shift in the number of people living on farms, as well as those who wanted to work on them, and labor issues started to emerge.

The trend toward livestock specialization—such as raising solely beef, dairy, swine, or sheep—accelerated during the 1970s and 1980s to the point where today few large farms raise multiple livestock species. If a farming enterprise does own more than one species, it typically has separate units on separate farms and the animals do not commingle.

Because of low input costs, sheep raising can be a viable option for the acreage you have available. Sheep can be raised outdoors in fields or pastures where they can roam

in all kinds of weather and thrive. Raising sheep offers a variety of options because of the very nature of the animal. You can enter and exit a small-scale sheep-raising enterprise quickly with only a short-term commitment needed from birth to market. With minimal investment, you can establish a small flock of two or three ewes, raise their lambs to sell quickly or to market weight, sell the ewes, and then exit within several months. If you wish to return at a later date, you can re-enter the sheep business with the purchase of pregnant ewes.

It will be extremely difficult to make any kind of substantial profit on two or three ewes and their resulting lambs, but this may not be the sole reason you want to raise sheep. If profit is your motive, you should consider raising a much larger number of sheep, where the numbers will be more advantageous to you. You can raise sheep for a larger market, especially if you decide

13

Your road to the future can include raising sheep. You do not need a large flock to derive enjoyment from them or the satisfaction of working with gentle animals. *Bill and Liz South*

to cultivate a niche market that attracts customers who appreciate the meat products and the way in which the sheep were raised.

SHEEP ARE NOT GOATS

Some people confuse sheep with goats, and although they may have some similarities, they are easily distinguished from each other. Sheep and goats are two distinct species. Sheep have fifty-four chromosomes and goats have sixty, which prevents successful crossbreeding between them.

There are two easy ways to determine whether an animal is a sheep or a goat: observe its tail and look for a beard. A goat's tail will rise and be held upward while a sheep's tail, if still attached, will hang down. This downward position is the main reason sheep tails are often docked, or removed, for health and sanitary reasons.

Goats have a unique divided upper lip and a beard hanging down from underneath their jaw, while sheep do not. Sheep breeds are often naturally polled (no horns) either in both sexes or just in the female, while naturally polled goats are rare. Males of the two species differ in that buck goats acquire a unique and strong odor during the rut (seasonal breeding period), but sheep rams do not.

In diet selection and feeding behavior, there are major differences between sheep and goats. Sheep are grazers, which means they eat grasses, usually on pasture, and eat close to the ground. Goats are sometimes termed browsers and prefer to eat leaves, shrubs, vines, branches and twigs, or weeds. Goats are more agile than sheep and will stand on their hind legs to reach vegetation, while sheep are not prone to do so.

From a behavioral standpoint, goats tend to be independent of one another, while sheep often exhibit a flocking instinct and crowd together in the presence of a strange animal or person. When a sheep is separated from the rest of its flock, it will typically become agitated until reunited. This flocking instinct provides their best defense against predators.

OTHER BENEFITS AND CHARACTERISTICS

Sheep have been used to develop products used in human physiological interventions such as arteriovenous shunts, which are devices that allow patients with kidney failure to be connected to a dialysis machine for long-term treatment. Scientists have also used sheep to develop models for decreasing the effects of osteoporosis in humans. Because of the similarities between sheep and human bone and joint structure and bone regeneration, sheep have been used to study bone healing and test synthetic bone replacements.

Several behavioral studies have concluded that sheep have an amazing ability for facial recognition and that some can remember the faces of as many as fifty other sheep for up to two years; they may even recognize a familiar human face. An Australian study suggested that sheep have excellent spatial memory and are able to learn and improve

A long wool breed such as the Wensleydale can have several color variations within its population. This makes them attractive because the natural wool color does not need to be dyed. *Cynthia Allen*

their performance. Perhaps this is connected to the construction of their head. Being a prey animal, sheep of the past required excellent senses to enhance their chances for survival in the wild, and these have been passed down through the succeeding domesticated generations.

Sheep depend heavily on their vision and have horizontal slit-shaped pupils. They have excellent peripheral vision, with visual fields ranging between 270 to 320 degrees. This allows them to see behind themselves without turning their heads. They have poor depth perception and may balk at shadows or dips in terrain, which are considerations when working with them in lanes and pastures.

Sheep have good hearing and, like most other farm animals, they can become frightened in the presence of barking dogs or other sudden sharp noises. Their excellent sense of smell alerts them to predators, helps rams locate ewes in heat, and allows mothers to locate their lambs. Touch is also an important sensory stimulus for sheep, particularly as an interaction between lambs and their mothers. Lambs seek bodily contact with their mothers, and the ewes' response provides a stimulus for milk letdown.

SHEEP NOTORIETY

Sheep have appeared in many forms of media production including movies such as *Babe*, literature such as George Orwell's *Animal Farm*, and nursery rhymes such as *Baa Baa Black Sheep* and *Mary Had a Little Lamb*. Sheep have been key symbols in fables and male sheep are often used as symbols of virility and power.

Most people are familiar with the term "black sheep," used to describe an individual who is perceived as odd or a disreputable member of a group or family. The term derives its meaning from the recessive gene for color that allows a black lamb to be born in an entirely white flock. This color was not as commercially viable as white wool and was considered undesirable by shepherds.

Perhaps the most famous sheep in the world, aside from Mary's little lamb, was Dolly, a Finnish Dorset sheep that was the world's first mammal to be cloned from an adult somatic cell. She was born in 1996 and resulted from the contributions of three mothers: her genetic mother contributed her DNA, a second ewe provided the egg into which the DNA was injected, and the third ewe carried the resulting cloned embryo and gave birth to Dolly.

Lambs quickly acclimate themselves to their surroundings and learn to eat grass very early in life. Lambs are your future and proper care and nutrition will provide you a new crop each year. *Gary Jennings*

PRODUCTS DERIVED FROM SHEEP

Meat and milk from sheep provide nutrition for the human diet. Many byproducts, such as hides, fats, intestines, glands, and bones, are used to make numerous other products utilized in our daily lives.

Meat is the most important product derived from raising sheep. Most producers raise sheep to sell meat by marketing through livestock auction markets or by direct marketing to customers. Lamb is the meat from a sheep that is less than one year old and mutton is from a sheep older than one year of age.

Wool was one of the first textiles and allowed early sheep growers to amass great wealth. Wool is used in clothing such as socks, sweaters, and suits. In more recent years, synthetic fabrics have replaced much of the need for wool, which has depressed prices to the point that in some areas the cost of shearing is greater than the possible profit from the fleece. This makes it almost

impossible to raise sheep only for the wool without the assistance of farm subsidies.

Perhaps the most recognizable product derived from sheep is lanolin, the waterproof, fatty substance found naturally in raw wool. It is a complex mixture of alcohols, esters, and fatty acids and is recovered during the scouring process. Raw wool can contain from 10 to 25 percent lanolin, and when processed it is used in such products as adhesive tape, printing inks, motor oils, and other lubricants. Lanolin can also be found in lipsticks, lotions, shampoos, and hair conditioners.

Sheep hides and skins are treated in a process called tanning after they are removed from the carcasses. The hides and skins are used in making leather goods, and a small number of skins are preserved and sold as sheepskins, often with the wool still attached.

Intestines of sheep are used for making violin strings and surgical sutures. Packers and processors get more usable byproducts from sheep than from other meat

Sheep grazing in your pastures can create a pleasant sight. They grow into animals that can be marketed, sold as breeding stock, used for wool, and milk production for making cheese. *Nancy Hann*

animals. At least sixty-seven byproducts are derived from sheep. For medicinal purposes, the adrenal glands produce epinephrine (adrenaline), which is a powerful heart stimulant. It requires 100,000 lamb adrenal glands to produce one pound of epinephrine.

Sheep cheese comprises about 1.5 percent of the world's cheese, including some of the most famous, such as Roquefort, feta, and ricotta. Sheep's milk can be processed into butter, yogurt, and ice cream.

EMERGING TRENDS MAY HELP SHEEP RAISING

As the pressure increases to secure sufficient energy sources to meet consumer demand, it is likely to affect agricultural production. Energy issues continue to make news as the world market price of petroleum products has risen and fluctuated for any number of reasons. Ethanol production has captured headlines because of its ability to stretch existing gasoline supplies. Its production, however, requires a vast amount of corn.

What does this have to do with raising sheep? The answer has both short- and long-term implications. In the short term, corn prices will likely rise overall and will impact expenses for those who purchase corn for raising sheep.

In the long term, this may benefit small-scale enterprises. Because sheep can graze and grow on grasses and pastures, hay or silage may be the only needed nutrition to grow to target market weights and may eliminate the need for expensive grain rations. By using feedstuffs that may be readily available on your farm and developing your own market, you may be able to circumvent these higher costs and still maintain a degree of profit.

Increases in corn prices affect large livestock enterprises, whether they specialize in pigs, dairy, or beef. Such increases may cause large livestock enterprises to become much less profitable and may affect total production levels, as farmers back away from feeding high levels of corn. This, in turn, may present an opportunity for those who wish to raise sheep on a smaller scale to have both a profitable and satisfying business.

Sheep are self-sufficient in many ways. Whichever breed you choose, you will need to learn how to care for your sheep, feed them, handle problems that arise, and how to market end products. *Shannon Phifer*

GETTING STARTED

If you own a farm, regardless of its size, you already have one of the most important assets required for raising sheep: land. If you are new to the farm real estate market, there are several issues to consider when purchasing farmland, including the location of the farm, soil type, house or dwelling, buildings available, and a number of other intrinsic factors such as schools, social outlets, and a sense of community.

How you handle these options may depend upon your financial situation, inclinations toward farming, and level of involvement in the prospective farm. If you already live on a farm but do not have any animals, you may decide that raising sheep is a viable option.

Farm ownership is not the only avenue for living on a farm. Renting a farm may be sufficient to achieve your goals. Whichever route you decide to take, there are many sources of information and advice available that can help you sort through all your considerations to arrive at a decision that is financially and emotionally satisfying. Advice for purchasing or renting an available farm can come from an agriculture lending group or bank, a county agricultural extension office, or private professional services that specialize in farm purchases and setting up farming enterprises. You can do much of the initial research on your own by contacting real estate agents about the availability of farms for sale or rent or by visiting properties on your own in locales where you may want to live.

Proper ewe and ram selection is critical to the success of a commercial sheep enterprise. Although crossbred ewes are more readily available and are less expensive than purebred ewes, purchasing purebred ewes can provide options such as developing a seedstock flock for other producers. *Jan and Coby Schilder*

PARTS OF A SHEEP

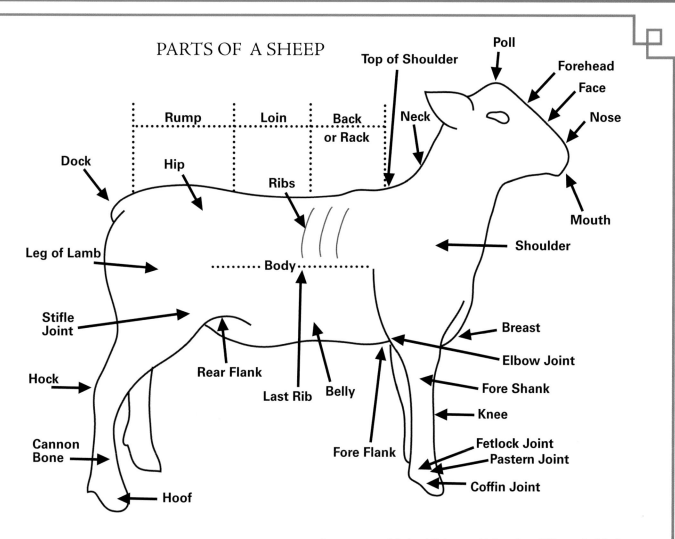

Department of Animal Sciences, University of Wisconsin-Madison

Line drawing by Lindsay Haas

Purchasing a farm is not the same as renting. If you purchase a farm, you are generally purchasing a business because there are financial considerations whether you work the land, rent it to another party, or leave it lie fallow. The financial obligations that come with purchased land, such as real estate taxes, make it almost necessary that you initiate a usage plan for your acreage. If you have off-farm employment, raising sheep is an excellent alternative for your farm because of the limited time involvement necessary to provide good management.

LOCATION AND SOCIAL CONSIDERATIONS

Property location and the services available may be important factors when deciding where to buy. Living in a rural area is not the end of the world; however, there are some significant geographic differences between rural and urban settings. Living on a farm does not necessarily exclude you and your family from the conveniences or services available in urban areas. There just happens to be a greater distance to access them.

In most cases, it is likely your farm will also be your home. It is important to assess whether the house or dwelling meets your family's requirements, both now and in the future. If you have a young family, it may be important to be located close to schools, doctors, or transportation systems. The availability of professional veterinary services may be important if your experience with managing animals is limited. If community activities are important, you can visit the area's chamber of commerce, which provides information about local events and activities held throughout the year. You may also want to look at the opportunity for alternative or off-farm income or employment.

If you have little experience with sheep, consider starting with a small flock of ewes. Before you buy sheep, read about all aspects of sheep production and consult with farmers who are currently raising sheep or with your county agricultural extension agent.

PHYSICAL CONSIDERATIONS

You do not need a large farm to raise sheep. The goals you set will determine the size of the farm you will need. How much land you will need to feed your sheep throughout the year will be a direct result of how many sheep you plan to raise, as well as the climate in which you choose to live. In many parts of the upper tier of the United States, grass grows abundantly if given a chance. Drier, more arid areas of the country, such as the Southwest and many parts of Texas, cannot produce vegetation comparable to their northern counterparts without extensive use of irrigation.

A general rule for determining the number of sheep you can have on any given land area is to figure that three ewes and their lambs require roughly one acre per year. This figure will need to increase if you live in a drier climate, but it generally does not get reduced, even if you live in a more temperate climate. If you choose to raise ten ewes and their lambs, then you would need an available land base for feed of about four acres or more.

If you already own or rent a farm, take the number of ewes available for grazing, pasture, or haymaking and divide that by the 3 acres to arrive at the number of ewes your farm can sustain. For example, 50 acres available for grass production would allow you to raise roughly 150 ewes with their lambs.

Your ability to raise sheep on farmland will also be influenced by other factors such as soil type and fertility. These may be important considerations to you because they also may be tied to property values. Soil type influences the crops raised and their durability in extremely wet conditions or during a drought or extended dry spell. Heavy soils sustain crops better in dry conditions, while lighter, sandier soils do not. On the other hand, sandier or lighter soils warm more quickly in the spring and provide better drainage in very wet conditions.

Another determining factor in your purchase may be the quality of the buildings and the number of improvements needed. Extensive building renovations require finances that could be directed toward farm operating expenses. Yet the need for improvements may lower the purchase price and be an attractive option.

Whether you purchase or rent a farm, it is necessary to fully understand its boundaries. Walking the fences provides you with an idea of how much land there is, as well as information about the condition of the fences, buildings, soil, and other aspects of the property, such as the suitability to pasture-raise sheep.

Prior to signing a purchase agreement, it is important to determine the presence of contaminants or residues that could affect the health of your family or livestock. Conducting a water test is a good idea and may be included in the purchase agreement. Underground fuel storage on farms has been banned in most states. Old stor-

age tanks may still be present and will need to be dug up and removed. Be sure to address the issue of underground storage tanks prior to signing a purchase agreement.

BUYING YOUR FIRST SHEEP

There are several ways to purchase sheep. Have a plan before you start. No matter which livestock species you're going to purchase, buying animals always contains a certain amount of risk. You can lower this risk by considering several factors when purchasing animals, including their overall physical condition, health, mobility, and knowing their source. These may not be the only criteria you use in your considerations, but they will provide a foundation for selection.

One of the first questions you should ask yourself before buying any sheep is what are your plans for them? Do you want to fatten them for food on your table or for marketing purposes? Do you want to use their wool

University sheep flocks offer opportunities for beginning sheep growers to study methods being used, purchase quality seedstock, and learn from experienced shepherds. A visit to a university sheep research station can be a very good learning experience. *Department of Animal Sciences, University of Wisconsin-Madison*

Start-up or capital costs are significant factors that must be considered when deciding on a farm purchase. These are typically paid for over several years. Farm land and buildings are the most expensive items for a farming enterprise.

for spinning or knitting? Do you wish to make sheep milk cheese? Or do you want them for scenery and as pets? It is best to answer these questions prior to making any purchase.

Regardless of your plans or the breed you choose, there are basic considerations regarding the health of the animals you select that will impact your long-term success. Although they are closely linked, physical condition and overall health are separate issues when

selecting sheep. A healthy animal typically is an aggressive eater and alert to its surroundings. The quickest way to determine the health of a sheep is simply to look at it. Does the animal appear to be alert? Does it have clear, dry eyes? Is it breathing normally? Does it move around easily?

Sheep that appear thin or emaciated should be avoided entirely, no matter the price, because this may indicate serious health problems. Avoid sheep with a dull

Your entire family can become involved with your sheep growing business. Working with animals can teach lessons such as discipline, dedication, and attention to detail, which can enhance career choices. *Jan and Coby Schilder*

It is essential to understand the boundaries of the farm you want to purchase. Exterior barriers will keep your sheep from entering your neighbor's property and keep theirs from entering your property.

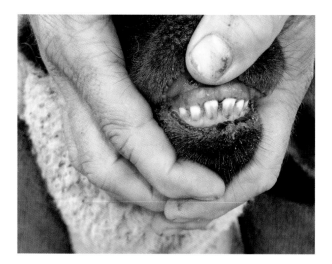

Sheep that are a year old should have eight teeth slightly protruding from the bottom gum and no teeth on the top pad. As sheep age during the first year, the first two middle incisor teeth will grow noticeably longer. In year two, the teeth on either side will grow. During year three, teeth will continue to grow and fill the mouth. A full compliment of eight incisor teeth should have grown in by the time the sheep is four years old.

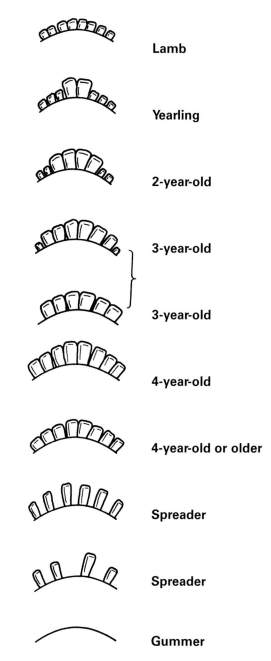

Lamb

Yearling

2-year-old

3-year-old

3-year-old

4-year-old

4-year-old or older

Spreader

Spreader

Gummer

The approximate age of a sheep can be determined by examining their teeth. Sheep have twenty temporary teeth but have thirty-two permanent teeth. Eight of these are incisor teeth in the front of the lower jaw, which reveal the sheep's age. This chart shows examples of differences in incisor teeth as sheep advance in age. *University of Wyoming*

appearance, discharges from their eyes or nose, breathing abnormalities, listlessness, or anything else that strikes you as abnormal.

Mobility is one health issue that can be visually observed. Any sheep, regardless of age, should have the ability to move about freely with no leg, joint, or feet problems. Avoid animals that have swollen joints, long toes, or malformed feet. Sheep exhibiting these conditions will not last long on your farm. If possible, try to observe the feces or manure droppings of the animals being considered. A normal, healthy sheep will have feces that resemble rabbit droppings like small balls that are not stuck together in large sausage shapes or in a runny consistency. Anything other than a normal size and shape could indicate an internal worm infestation.

You can get an approximate age range of ewes by examining their teeth. At birth, lambs have eight milk teeth, or temporary incisors, arranged in four pairs on the lower jaw. At approximately one year of age, the central pair of temporary incisor teeth is shed and is replaced by permanent teeth. At two years, the second pair of milk teeth is replaced by another pair of permanent incisors. At three and four years, the third and fourth pairs of permanent teeth appear, and at four years of age the sheep has a full set of teeth. A mature ewe

Regardless of the age of the sheep you acquire, you should purchase only healthy animals for your flock. Healthy sheep grow better, require less handling, and are less likely to spread diseases. *Cynthia Allen*

of four years of age should have a full set of eight permanent front teeth (incisors) along her lower jaw. Sheep have no teeth along the front of their top jaw—only a hard dental pad. Teeth may be well worn but they should all be there, firmly fitted and even. Lack of a suitable number of teeth may be a hindrance for the animal to eat enough to stay healthy, as well as indicative of being much older than claimed. Look for square jaws. A sheep's lower jaw should meet squarely with the upper pad. Animals with either overshot (short lower jaw, known as "parrot mouthed") or undershot (long lower jaw, known as "monkey mouthed") jaws should be avoided.

If you are not sure of your own expertise at identifying problems, hire a veterinarian or someone with sheep experience to go along to look at the animals you are considering for purchase. This will be money well spent and will help avoid problems.

Knowing the source of your sheep may alleviate many worries about their health. If you purchase sheep from a private owner, take a look around the farmstead. If it is well kept and clean, it is likely the farmer pays attention to the details of his or her sheep. Observing the attitude of the person selling the animals can often provide clues as to their treatment and care. Having the satisfaction of buying in a pleasant surrounding from a caring farmer can ease your concerns about the sheep you buy.

If you have little or no experience with raising farm animals, you may want to start with a pair of ewes. This

Buying a ram is a significant investment for one animal, and it also is a determining influence in your future flock. Use good judgment in acquiring a sound ram. The fastest way to bring new genetics into your flock is with a superior ram.
Jan and Coby Schilder

A good ewe can cost between $100 to $250, while the price for a ram may range between $250 to $450, depending on breed, availability, and other market factors. Your investment in quality animals will be repaid in lambs that quickly reach target market weights.
Nikole Riesland-Haumont

minimizes your initial investment, requires less labor, and allows you to familiarize yourself with the sheep-raising process. As you gain experience, confidence, and expertise, it will be easier to plan for more sheep.

START-UP ECONOMICS

You can calculate potential starting costs by using certain criteria and assumptions that would be valid for the area in which you live. For example, buying pregnant ewes is a quick way to increase your flock numbers. At this writing, pregnant ewes can be purchased for $100 to $250 each. Breeding-age ewes will cost about $80 to $90 each. Weaned lambs old enough to be considered for raising to maturity will cost roughly $1 per pound. Rams may cost $250 to $450 each. These prices reflect strictly commercial sheep and not registered animals or heritage breeds that may cost more because of their value as breeding stock or their diminished availability.

The largest expense you will incur while raising and keeping sheep is feed costs. This expense will be greatly affected by whether you grow your own feed or purchase all or most of it. Because sheep are ruminants, you will have several options in developing feed rations including a mixture of grasses, pastures, and grains in formulations that can keep feed costs to a minimum. Lambs require more specialized growing rations, which may be of your own design or purchased as a premixed supplement.

A mature ewe will typically consume one to two pounds of grain, along with two to four pounds of hay per day for maintenance and early gestation. During late gestation and lactation this can increase to four to seven pounds of hay per day. On a yearly basis, one ewe can consume between 400 to 800 pounds of grain and between 1,000 to 2,000 pounds of hay. Costs for raising lambs to market weight (115 to 140 pounds) can also be calculated. Feeder lambs will eat between two to three pounds of grain per day and up to four pounds of hay. Using available market values for grain and hay in your area will allow you to quickly calculate potential costs for each animal.

WHERE TO BUY SHEEP

Sheep can be purchased privately from another farmer or at public sales that include feeder lamb sales, auction barns, or sales on the Internet. Each venue has its advantages and disadvantages. A public auction is where

Rams that grow quickly and exhibit superior physical characteristics will typically pass these qualities to their offspring. Healthy rams will exhibit more of a willingness to breed ewes, which will result in higher conception rates. *Nikole Riesland-Haumont*

Check all the feet and toes of each sheep you purchase. The nails or toes should not be overgrown or malformed. With good visual and manual inspection, you should be able to detect any problems. Toes should be well-formed and the skin between the hooves should be intact and show no signs of swelling or redness.

anyone can bid on animals to purchase. The winner pays the highest bid price. You will be expected to present a good check after the sale. A number of sheep breeders hold public auctions, sometimes in conjunction with state, regional, or national shows. These are good venues for purchasing animals because the quality is generally higher than at a typical auction barn sale, and established breeders typically offer stock from their better animals.

One advantage of buying at a public auction is the established conditions under which the sheep are sold. Sometimes guarantees regarding the animals are stated at the beginning of the sale.

Buying animals at a breeder auction has an advantage in that the price for those animals is determined by other bidders. This can provide a reasonable assessment of the worth of those animals by other, more experienced sheep growers and affirm your judgment of the animals. The price may be more than you want to pay, however, and you may go home empty-handed. If you have chosen

a particular breed, attending a breeder sale can introduce you to many other farmers who may have animals available for sale. Public sales can be good social events where you can meet other like-minded people, and the contacts you establish can help you develop long-term friendships and business acquaintances.

Private sales allow you to acquire all your sheep at one place. This will generally reduce the risk of exposure to other sheep from different farms, which is a health consideration. Although you may pay a price determined by the owner, you also may be able to negotiate a better price if you purchase several animals at one time. An established sheep producer's reputation is important, and most will try to accommodate a buyer's concerns if something should go wrong later.

BUYING A RAM

Buying a ram is different from buying feeder lambs or ewes for breeding. The growth rate of your lambs will be determined by two things: your feeding program and the genetic inheritance from their sire and dam. The fastest way to bring in new genetics without replacing the entire female population is to use a superior ram. Whether you buy a ram for breeding purposes or raise one from your own flock depends on your situation. Some cross-breeding programs use a rotational plan that produces replacement females so only rams need to be brought into the flock. This also keeps the flock exposure to outside health concerns to a minimum.

Growth rates and carcass traits may be a consideration whether you are selecting for the general market or for a niche market. Try to select a ram to improve both. This may not be possible in some heritage breeds where the genetic pool is limited and genetic advances for carcass and growth traits may be slower.

If you choose to purchase a ram for breeding your flock, it is important that this purchase be made several months prior to the breeding season. Rams need time to get acclimated to their new surroundings and the feed rations they will be consuming during the breeding season.

There are several physical considerations to keep in mind when purchasing a ram. Many are similar to the general considerations for purchasing females: healthy

feet and legs for mobility, clear eyes and lungs, lack of internal parasites, and good teeth. Other considerations for rams include well-developed sex organs, no hernias, full hindquarters that suggest a good meat-producing ability, and a good fleece, if that is your market.

Well-grown males that achieve a fast market weight will generally help accelerate the growth rates of his off-spring. Rams that achieve target weights earlier than others will most likely have a positive effect on your flock and will be best positioned for sexual maturity.

Proper ram management will positively affect the reproductive efficiency of your flock because the ram must be healthy to effectively impregnate the females. Failure to do so will result in lost pregnancy time and fewer lambs born on schedule, which will ultimately affect the total number of lambs born and flock prof-itability. The libido or willingness of the ram to breed

Mobility is easily observed. When buying sheep you should look for ease of movement in their legs, joints, and feet. Avoid buying sheep with observable lameness, swollen joints, long toes, or malformed feet.

ewes is highly variable among rams and can have a major impact on sheep production, particularly if a single sire is used in the flock. Libido is affected by underfeeding or overfeeding where the ram is underweight or too fat to become interested. Age and disease conditions such as arthritis may affect a ram's performance. The easiest way to determine a ram's mating behavior is to observe his performance as he is exposed to ewes. If he fails to show any interest to ewes in heat, he should be replaced as soon as possible.

A good healthy ram can breed twenty-five to thirty ewes and can usually mate three to four ewes per day

Alert, animated, and hungry sheep are a good indication of normal growth and health. Avoid sheep that appear listless and avoid mixing in with the rest of the flock.

without any noticeable effects. During the breeding season, a ram may lose up to 15 percent of his body weight. It's important that he is in top condition prior to use. A good feed ration will keep your ram in good physical condition during the breeding season and help ensure good conception rates.

Several management practices enhance a ram's performance including shearing, treating for internal parasites, trimming feet if necessary, and a flock vaccination program. It is important to maintain a cool, shady

place for the ram to live. An elevated body temperature, whether from heat or an infection, can cause infertility. Semen quality is affected at 80 degrees Fahrenheit and seriously damaged at 90 degrees or above. If you plan for a January or February lambing schedule then your ram will be servicing your ewes during August and September, which are typically high-temperature months in most of the United States. Therefore, in most areas of the country, care will need to be taken to prevent excessive heat from affecting your ram.

While horns may be attractive to look at, they can be a danger to you or your children. Carefully consider the advantages or disadvantages of raising a horned breed versus a polled (non-horn) breed before buying sheep. *Cynthia Allen*

To ensure that your ram is ready for service, you may want to have your local veterinarian perform an extensive physical examination prior to the breeding season. Some of the conditions the veterinarian may be able ascertain will be cryptorchidism, which is the failure of one or two testes to descend into the scrotum. Cryptorchids are undesirable breeding animals and should be eliminated as sires for your flock. Pizzle, or sheath rot, is an infection in the sheath area and can affect breeding activity. A qualified veterinarian will be able to determine whether your ram has either of these two conditions and may be able to provide a semen evaluation of your ram to determine that there are no sperm problems that will affect breeding efficiency.

ANIMAL IDENTIFICATION

If you are purchasing ewes, rams, or feeder lambs and having them transported to your farm, it is important that you properly identify those animals at the purchase time. This will eliminate any potential problems with getting the wrong animals. Proper identification is an important management tool that provides many benefits as you progress in your sheep-raising program.

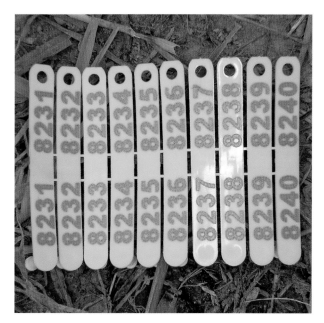

Proper identification of individual members of your flock starts when they are lambs. This aids in recordkeeping and management of your flock. Many states now require a farm premise identification number that is unique to each farm with animals. Contact your county agricultural extension office to learn more about this program.

CHAPTER 3

• •

PRODUCTION SYSTEMS

There are two different models under which to raise sheep—production and economic—and each model can be a part of the other or have further subdivisions. The production model involves organic, sustainable, or conventional protocols. The economic models involve the type of day-to-day program you wish to use, such as market lambs, feeder lambs, wether production, purebred stock, meat production, milk production, or wool production.

Organic and sustainable systems are gaining popularity for several reasons. The most prevalent are the increased markets for products from these systems and the increased value those products bring to the marketplace. There are significant differences between organic and sustainable farming and conventional farming practices. You should consider each practice carefully before deciding which one to choose.

The economics of your sheep-raising program may be dictated by the model you choose. Sheep can be sold as feeder lambs or raised to market weight. You might raise purebred stock for breeding purposes. Sheep can be raised for meat, milk for making cheese, or wool. Having a goal at the beginning of your program will make better use of your resources and help you make the best decisions for your overall program.

CONSUMER TASTES

Shifts in consumer tastes and a higher demand for foods that come from organic and sustainable farms are making these production systems attractive to farmers. Conventional farms use intensive planting and harvesting of crops typically aided by crop chemicals and commercial fertilizers to increase production and generate higher profits. The current shift in consumer demand for food

Regardless of the production model you choose, good management, feeding, and day-to-day care is essential to keep sheep healthy so they can quickly reach target growth rates.
Jan and Coby Schilder

The range of marketable products from sheep makes them a good choice for small-scale farming. Sheep can be raised for market or for breeding stock, or they can be sold as feeder lambs to other growers. They can be raised by conventional practices or using organic and sustainable production. *Cynthia Allen*

products produced without chemicals or intensive farming systems has been driven by the perceived benefits from more traditionally raised animals. This consumer shift, which cultivates the image of healthier foods, has brought more money to organic and sustainable markets. It has allowed farmers to utilize practices that may be more in tune with their personal ethics and nature's harmony. Organic and sustainable farming can be a good fit for small-scale farming.

ORGANIC PRODUCTION

Organic farming emphasizes management practices over volume practices; considers regional conditions that require systems to be adapted locally; and supports principles, such as environmental stewardship. Organic farming is defined as a holistic, ecologically balanced approach to farming that supports, encourages, and sustains the natural processes of the soil and the animals. Those involved with organic farming have a perspective

that reaches beyond the present and requires a long-term commitment to their program.

With organic farming, the whole ecosystem of the farm is incorporated into the production of animals and crops. It seeks to obtain the greatest contributions from on-farm resources such as animal manures, composts, and green manures for soil fertility and to eliminate external additives, especially synthetic chemicals.

While on-farm resources are an essential part of the operation, organic farming is flexible enough so that if sufficient quantities of materials cannot be produced on the farm, off-farm nutrients such as natural fertilizers, mineral powders, fortified composts, and plant meals from approved organic sources can be applied without risking certification concerns.

Organic farming promotes the health of the soil by encouraging a diversity of microbes and bacterial activity. This in turn enhances the growth of plants and grasses deemed healthier for the grazing animal.

ORGANIC CERTIFICATION

If you pursue organic production, you will need to become certified. Certification is required for producing an organic product that is packaged and labeled as such for sale. The term organic is defined by law in the United States and the commercial use of organic terminology is regulated by the government. While the process to become certified can seem extensive, the result is that products produced under certification are authenticated.

Lamb sales can make up a large percentage of a farmer's income from sheep production. Spring and winter lambing are the two most common production systems used. *Shannon Phifer*

Choosing to farm organically will require a change from traditional farming practices. Grains may be fed to your animals, but under organic rules it must come only from sources that produce it organically in order to maintain your certification. Specialized organic markets must be found to purchase feedstuffs used on your farm and for the sale of livestock to get the greatest benefit from your efforts.

There is a three-year transition period between using conventional farm practices and achieving organic certification. No chemicals of any kind can be used on the land or on your sheep during this transition period. Antibiotics cannot be used on animals except for lifesaving interventions, after which the animal can no longer be marketed as organic. Genetically modified organisms (GMOs) cannot be planted. The exclusion of insect- and pest-control products can make raising sheep seem more of a challenge, but it is being done in many different parts of the country with great success.

SUSTAINABLE FARMING

Sustainable farming is a goal rather than a specific production system. Its goal is to achieve a balance between what is taken out of the soil and what is returned to it without relying on outside products. The aim is for perpetual production.

Sustainable production principles for raising sheep include consideration of how you integrate your crop and animal systems. The key components are profitability, management of your animals, and stewardship of natural resources, such as water and soil.

Most adverse environmental consequences associated with your animals, including stocking rates, can be reduced or eliminated through proper grazing methods and pasture management so that limited amounts of feed must be brought in from outside sources. The long-term carrying capacity of your farm must take into account short- and long-term dry spells and reduce overgrazing on fragile areas of the farm.

Animal health is an important part of a sustainable farm because the health of your animals greatly affects reproductive performance and weight gains. Unhealthy livestock of any species waste feed and require additional labor. Sustainable sheep production relies on

Winter lambing generally occurs from late November through February. In most regions of the United States, covered facilities are needed during winter lambing to protect the ewes and their lambs. *Jan and Coby Schilder*

pasture-based production methods that lower the costs of mechanical harvesting because the sheep harvest the grass themselves.

Your farm can be sustained by developing a strong consumer market for your products. Becoming part of a coalition can be useful in promoting your products and educating the public in general. By educating more of the public, you will likely increase your market. Sustainable sheep production may be an option for you because it focuses on the long-term health of the environment, while making the farm economically viable and addressing social issues that you may consider important.

CONVENTIONAL FARMING

Many conventional farming practices are geared toward maximum production by using large amounts of outside resources to produce their products. Whether it is corn, soybeans, alfalfa, wheat, or any number of other crops, intensive farming practices have become the norm for many farms. These farms typically use chemically based products to control insects, pests, and weeds, and to promote rapid growth of crops and animals or to increase milk production. Hybrid seeds or GMOs are sometimes planted to increase yields. Highly specialized machinery does most of the work and the operator's feet may seldom touch the ground.

Conventional farming differs dramatically from organic and sustainable farming in that there is a heavy reliance on nonrenewable resources, such as fertilizers, gas, and diesel fuels. There is also the concern that some practices, such as excessive tilling, can lead to soil erosion that may cause long-term damage to the soil.

SHEEP PRODUCTION SYSTEMS

There are several production systems you can use when raising sheep: meat production, lamb feeding, wool production, dairying, and breeding stock. Some breeds will perform better in certain systems than others. You may want to research those differences prior to choosing a breed and setting your goals. The production practices you use will vary between these different systems.

MEAT PRODUCTION

Meat sheep producers typically raise slaughter lambs or feeder lambs for sale to market. Most sheep and lambs are bred for meat production in the United States and kept for the replication of lambs for meat, while dual breeds can be kept for both the production of lambs for meat and wool.

You can sell either feeder lambs or market lambs because these are the most easily replenished animals in your flock. Unless they need to be sold for a variety of reasons, including failure to breed or old age, selling mature ewes will deplete your breeding flock and it will take longer to replace them.

Feeder lambs are kept for the purpose of feeding for slaughter and typically range from 50 to 90 pounds. The highest demand is usually for 60- to 70-pound lambs. Market lambs are usually sold for immediate processing.

Spring lambing refers to lambs born from March through early May. This system allows the lambs to grow on milk, creep feeds, and summer pasture in order to be ready for market in late summer. Covered facilities are beneficial but are not essential with this system. *Nikole Riesland-Haumont*

Lambing in early spring allows for both barn and pasture lambing. Unassisted pasture lambing can be practiced but it will slightly increase lamb mortality percentages, particularly with multiple births. This loss percentage can be reduced by choosing a sheep breed known for its easy lambing traits, although it may still vary in individuals. *Nikole Riesland-Haumont*

The average market weight of lambs in the United States is about 130 pounds. The ethnic market, however, typically requires lambs that are much lighter in weight—about 100 pounds or less—but there is also a market for lambs of varying weights.

If you decide to raise sheep for meat, there are several factors that will determine your profitability, including percent lamb crop, lamb growth rates, and market prices. The percentage of lambs that are born and survive will be greatly influenced by your management at breeding and lambing time. Securing and using a fertile and vigorous ram will help achieve high pregnancy rates. Missed pregnancies will lower the number of lambs born during the year. Lamb growth rates will determine how fast you can get them off to market. This will be influenced by genetics and the rations you use. Market prices may be beyond your control, but studying market trends may help you find the appropriate time to send your lambs to market. If the lambs have reached their target weight, however, you may not have much control over the situation.

Lambing from September to November is referred to as fall lambing. It reduces the need for covered facilities, increases the utilization of high quality fall forages through lactating ewes, and provides producers more time to finish slaughter lambs for traditionally high winter and spring markets. *Department of Animal Sciences, University of Wisconsin-Madison*

FEEDER LAMBS

Depending on your location, a commercial feeder lamb or finishing lamb program may be a viable option for your farm. In colder climates of the United States, feeder lamb production is a seasonal enterprise while in warmer parts of the country it can be a year-round business.

With feeder lamb production, the lambs are purchased when they are between 50 to 90 pounds and are fed to a finished weight of between 100 to 130 pounds, depending on the target market. Feeder lambs can be raised on different diets including complete rations, whole grain rations, or forages. They can be successfully finished on pasture if there are sufficient grasses or a combination of pasture and grain. Lambs should be at least 60 to 65 pounds to effectively utilize a high-roughage diet, and the total digestible nutrients (TDN) must be 70 percent or greater to achieve finish. Whichever diet is used, finishing programs should take advantage of the resources available as lambs are capable of turning a wide variety of high-energy feedstuffs into growth.

If you decide to use a heavy grain ration for finishing feeder lambs, be sure to introduce grain slowly at first. Find out how the lambs were raised prior to their purchase. Transitioning to grain requires good management skills to avoid putting lambs off feed and to minimize incidents and losses due to acidosis.

WOOL AND HAIR PRODUCTION

Wool is the product most commonly associated with sheep. Although wool was the first commodity to be traded internationally, today its importance has dropped dramatically because of the emergence of synthetic fabrics.

Fall lambing is more difficult to accomplish because of the seasonality of breeding inherent in most breeds of sheep. By incorporating certain breeds, such as American Blackbelly, you can develop an accelerated lambing program because of their out-of-season breeding capabilities. *Department of Animal Sciences, University of Wisconsin-Madison*

Specializing in the production and sale of breeding stock is an option for sheep growers. Breeding stock sales can include ewes and rams, purebred registered animals, and commercial crossbred sheep. Customers for breeding stock may be other seedstock breeders or commercial sheep producers. *Department of Animal Sciences, University of Wisconsin-Madison*

Sheep dairying is a growing industry in the United States. Most of the U.S. sheep dairies are presently located in the Upper Midwest (Wisconsin and Minnesota), California, and the New England states. The East Friesian is one of the best United States breeds for milk production capacity.

Specialized milking equipment is required for sheep. Using a parlor system similar to the system used by dairy cows reduces labor and provides a fast and efficient method for milking sheep. Sheep quickly learn the parlor milking routine. *Department of Animal Sciences, University of Wisconsin-Madison*

Historically, the value of a sheep was in its wool, but today a sheep's value is in other products. There is a subsidy for wool but it generally does not provide enough income for a farmer to be able to use sheep solely for wool production. Selling wool in the commercial wool market has limited profit potential. Wool can be a valuable product if used in niche marketing. The difference may be ten- to fifteen-fold.

High quality wool can be sold to hand spinners. Quality is largely determined by feeding and housing, health care, and handling of the sheep and fleece once it is removed. To be of greater value to hand spinners, fleece should be skirted, which is the removal of the undesirable parts such as belly wool, stains, coarse wool, cotted wool, and short wool.

The most valuable wool produced is taken from the initial shearing of a year-old sheep. This is the finest quality produced during the sheep's life. As the sheep ages and its diet changes, the quality of wool, though it still may be high, decreases over time.

Hair sheep production may be one option that is attractive to you. It is estimated that about 10 percent of

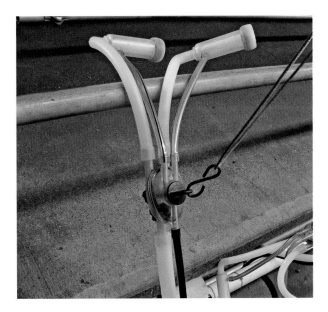

In a parlor milking system, sheep are milked from the rear with a twin-cup milking claw that is placed on the two teats. The vacuum system gently pulls the milk from the sheep's udder. The cups are removed after all the milk has been extracted. The sheep are released from their stalls when finished milking and return to their housing area.

Milk flows from the udder through a stainless steel pipe and drains into a glass receiving jar before it is pumped into a bulk tank. Cleanliness of the udders and equipment helps maintain sanitary conditions for producing quality sheep milk.

A flock of milking sheep is managed in much the same way as other flocks, although higher nutrition levels are required because of milk production. Ewes used for milk production typically lamb once a year. Lambing is the physical process that initiates lactation.

the world's sheep population is hair sheep. Although it is a small market in the United States, it is a growing one.

Hair sheep naturally shed their coats, which are a mixture of hair and wool fibers. They do not require shearing and tend to be more resistant to internal parasites than wool sheep. They tend to have many desirable reproductive characteristics such as early puberty, prolificacy, and different breeding cycles during the year. Hair sheep ewes can lamb on pasture, and the lambs grow well in a grass-based program. These advantages are presently influencing a growing market for hair sheep breeding stock.

SHEEP DAIRYING

A growing segment of the sheep industry produces sheep milk for cheesemaking. Sheep were milked long before cows, and while the dairy sheep industry is in its infancy in the United States, it is growing. At this writing, most of the sheep dairies are located in the Upper Midwest, California, and the New England states.

Most sheep milk is used to make cheese, and some is made into yogurt and ice cream. Any breed of sheep can be milked, but there are breeds that have better milk production than others. The East Friesian of Germany is one of the most popular dairy breeds in the United States, along with the Dorset and Polypay. The East Friesian can produce between 400 to 1,000 pounds of milk per lactation. The Dorset and Polypay can only produce 100 to 200 pounds per lactation. Crosses between the Dorset and East Friesian have shown an increased production level as high as 250 to 650 pounds of milk per lactation.

It is possible to milk sheep and still allow the lambs to nurse. One study has shown that leaving a lamb with an ewe for half the day and milking the ewe once per day during the first thirty days after lambing followed by twice-daily milking was the most profitable system because the ewes raised their lambs and still produced 85 percent as much milk as the ewes that were milked twice per day from shortly after lambing.

Dairy ewes have a significantly higher nutritional and water requirement than ewes being raised for meat or wool. Their requirements will depend on several factors including stage of lactation and lactation length. Milk production requires more nutrients because of the physical process of converting feed into fluid milk. Proteins, fats, and water are passed from the rumen to the mammary system to create milk. The length of lactation—the number of days between lambing during which a ewe is milked—will require maximum nutrition to meet the ewe's requirements of body maintenance, milk production, and preparing for the next lambing. A decrease in nutrition will result in lower milk production, loss of body weight, and less body condition at the time of the next lambing.

One advantage of sheep dairying is that dairy lambs are weaned much earlier than nondairy lambs. Dairy

A milking parlor with a pit allows the people doing the milking to stand even with the ewe's udders. This minimizes physical labor for the farmer and provides a clean environment for both the farmer and the sheep.

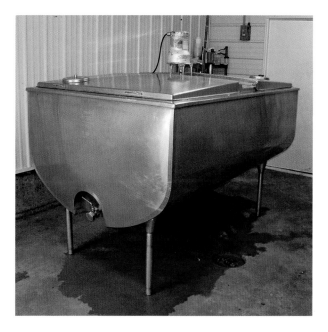

A stainless-steel bulk tank holds the milk after it comes through the pipeline. It has a cooling mechanism to keep the milk from spoiling before it is sold or processed into cheese.

lambs can be weaned at thirty days of age compared to their nondairy counterparts that average sixty to ninety days at weaning.

Sheep dairying requires milking facilities and equipment that must pass periodic state sanitation inspections.

While organic and sustainable production models are not the only ones available for you to consider, they are attracting more consumer interest and adaptation to farming programs because of higher prices received. You can take advantage of these markets by learning more about them from organizations involved with organic and sustainable production. *Nikole Riesland-Haumont*

Hair sheep production has seen an increase because of the declining value of wool relative to meat, along with the decreasing number of sheep shearers. It is estimated that approximately 10 percent of the world's sheep population is hair sheep and the U.S. population is over 3 percent. The American Blackbelly seen here is a hair sheep. *Department of Animal Sciences, University of Wisconsin-Madison*

The equipment must be stainless steel. Some producers use a milking parlor system in which a number of sheep can be milked at the same time by using multiple milking machine units. The milk needs to be cooled in a stainless-steel container called a milk tank before it can be sent for processing. It is also possible to freeze sheep milk for later processing; this does not decrease the milk's quality. Freezing the milk enables infrequent milk collection and allows farms located a great distance from a processing plant to produce sheep milk. Sheep milk used for a variety of artisan cheese products can be a niche market that leads to a profitable sheep enterprise. You may also harvest the meat and wool products of dairy sheep. More information about sheep dairying and the equipment needed is available from your county agricultural extension agent.

BREEDING STOCK

You may want to consider specializing in the production and sale of breeding stock as a sheep enterprise option. Breeding stock typically includes animals registered with a specific breed organization and includes both ewes and rams. Some crossbred animals may be included in breeding stock programs.

Recordkeeping is an important aspect of breed stock production, and there are computer programs available to help with your records. The National Sheep Improvement Program (NSIP) is a computerized performance record program that is useful for small producers and

One production model is to raise lambs and sell them when they reach feeder lamb market weights of between 50 to 90 pounds. Selling the lambs and retaining the most productive ewes will allow you to replenish your flock. *Department of Animal Sciences, University of Wisconsin-Madison*

allows for comparison of sheep from different flocks under different feeding and management programs.

Contact individual breed organizations for more information about their programs including consignment sales, production sales, and other breeders offering breed stock. While raising breeding stock is a more specialized business than some other production options, it can be a rewarding goal.

Above: Purchasing feeder lambs between 50 to 90 pounds and raising them to market weights between 100 to 130 pounds is another production system. It has several advantages including not having to feed ewes for the full year. *Steve Pinnow*

Left: Long wool breeds such as the Cotswold can provide additional income through specialized wool sales. Marketing fleeces directly to handspinners or by processing wool and marketing the value-added product can be another source of revenue for producers. *Sue Carey*

CHAPTER 4

BREEDS OF

SHEEP

There are more than fifty breeds of sheep in the United States and over one thousand breeds worldwide. About a dozen breeds in the United States are commercially important and have a wide diversity of use and purpose. Sheep breeds can be categorized several different ways to help you identify those that will

fit your needs or preferences. Your goal for raising sheep should be the primary consideration when deciding upon a breed or breed type. For example, if you wish to raise sheep for wool, your choice will be different than if you raise them primarily for meat products.

There are five accepted ways to categorize sheep breeds: commercial use, primary purpose, face color, type of fiber or coat they grow, and special traits. Another way to classify sheep is to identify them as crossbred, pure-bred, or registered; or as horned or polled. Similarities in production traits occur between different breeds, so it may be more useful to consider breed types rather than an individual breed. Breed types tend to share common characteristics and can usually be substituted for one another in a breeding program. There are many options available for choosing a breed, and understanding your end goal will help make your selection easier.

PRODUCTION PURPOSE

There are four categories for sheep when dividing them according to their commercial use. They are meat, wool, dairy, and dual-purpose. Although most sheep breeds

The Wensleydale is a large sheep with long-stapled, lustrous wool that falls in long ringlets almost to ground level. They have a bold, alert carriage and show considerable presence. This Black Wensleydale is a prime example. *Sherry Carlson*

can be considered dual-purpose, such as those used for wool and meat, some are triple-purpose (meat, milk, and wool). Most breeds generally excel in the production of one product, while other products are often secondary.

Meat breeds produce milk for their young, although it is generally not of sufficient quantity to consider them a breed for milk production. Sheep breeds available in the United States typically used for meat include Cheviot, Dorset (polled and horned), Hampshire, Montadale, North Country Cheviot, Oxford, Shropshire, Southdown, Suffolk, Texel, and Tunis.

Wool breeds can be divided into two groups: fine wool and long wool. Fine wool breeds include American Cormo, Booroola Merino, Debouillet, Delaine-Merino, and Rambouillet. Long wool breeds include Border Leicester, Coopworth, Cotswold, Lincoln, Perendale, Romney, and Wensleydale.

Dairy sheep breeds are those used mainly for milk production and cheesemaking, including East Friesian and Lacaune. Dual-purpose (meat and milk) breeds include the Border Leicester, Columbia, Coopworth, Polypay, Corriedale, and others.

COMMERCIAL USE

Commercial-use sheep breeds can also be categorized by whether they are more suitable as a ram or ewe breed. Ram breeds tend to excel in growth and carcass traits, while ewe breeds' strengths are reflected in maternal characteristics such as fertility, milking, and mothering ability.

FACE COLOR

Differentiation of sheep breeds can be made by face color. Sheep breeds with black or nonwhite faces, such as Suffolk, Hampshire, and Oxford, tend to exhibit better growth and carcass traits. White-faced breeds, such as Rambouillet, Polypay, and Targhee, generally have better maternal and wool traits.

WOOL OR COAT TYPE

The type of fibers sheep grow (fine, medium, long, or carpet) or the type of coat they have (wool or hair) is the most common way of categorizing sheep. All sheep grow both wool and hair fibers and the predominance of one over the other categorizes breeds. Hair breeds,

Dairy sheep breeds are used primarily to produce milk that can be processed into cheese. The East Friesian produces the highest volume of milk per year of any sheep milk breed. Shearing ewes at lambing makes the milking process easier.

The Border Leicester is considered a dual-purpose breed that can be used equally for meat or milk. Border Leicesters make very good mothers. They are descended from the Cheviot and Leicester breeds.

including Barbados Blackbelly and Romanov, have more hair fibers than woolly fibers and shed their coats annually, which eliminates the need for shearing. Wooled breeds have more woolly fibers and need to be sheared, usually once a year. Hair sheep are gaining in popularity because they do not require shearing or docking.

Fine-wool sheep, such as Debouillet, grow fleeces that are the shortest in length, are the smallest in fiber diameter, and contain the most lanolin. Fine wool tends to have the most valuable market price because it is used to make the highest-quality wool garments. Fine-wool sheep are well adapted to arid regions and comprise the most numerous sheep flock in the United States.

Long-wool sheep have long fibers that are larger in diameter than fine-wool sheep. They have less lanolin and yield a cleaner fiber. Long-wooled breeds tend to be favored by hand spinners and weavers. They are better adapted to wet, cooler climates than most long-wooled

breeds. The Cotswold, Lincoln, and Border Leicester breeds are descended from sheep on the British Isles.

Medium-wool sheep have length and fiber diameters that are between fine and long. Most of the breeds used for meat products grow medium wool.

Carpet wool is even longer and coarser than long wool, but carpet wool comprises only a small percentage of total U.S. wool production.

SPECIAL TRAITS

Another way to categorize sheep is by their special traits, such as type of tail and litter size. Fat-tailed or fat-rumped breeds are well adapted to arid regions. The Karakul breed is a fat-tailed breed, and the Tunis and Dorper have fat-tail origins. Some breeds, such as Romanov, Booroola Merino, and Finnsheep, are more prolific and are known for the birth of large litter sizes. It is very desirable for ewes to have twins.

HERITAGE BREEDS

Preserving rare and heritage breeds of livestock has become increasingly important to preserve species diversity. As the industrialization of farms and farming have evolved, production methods have favored fewer breeds exhibiting certain genetic traits that adapt well to intensive farming practices. The loss of unique breeds over time has meant the loss of the valuable genetic diversity they possessed. A growing number of farmers are interested in preserving threatened breeds. By choosing to raise a heritage breed, you can contribute to sustaining a unique population of sheep for the future.

The American Livestock Breed Conservancy (ALBC) has identified several sheep breeds as being threatened with extinction. Breeds listed as threatened include Cotswold, Jacob, Karakul, Leicester Longwoold, Navajo-Churro, and St. Croix. These breeds may be ones to consider on your farm.

HORNED OR POLLED

Depending on their breed, sex, and genetics, sheep may or may not have horns. The advantages of raising polled sheep—those born without horns—may outweigh raising those with horns. On a typical sheep farm, polled sheep make handling easier and safer for family members. Horned rams may be more difficult and dangerous to have around, especially if you have young children. Horned sheep also may get their heads stuck in fences, feeders, and equipment, and require more care and attention. Some heritage breeds, such as the Jacob, have horns that make them an attractive addition to a flock. Their unique four horns make them stand out in a flock.

The Romanov is a prolific breed that is known for multiple births. They are an early maturing breed that has an easy time lambing. *Don and Janice Kirts*

Jacob sheep have multiple horns (as many as seven) with a medium fine fleece. They have no outer coat, which sets them apart from other primitive breeds that are double-coated. They are slender-boned and provide a flavorful, lean carcass with little external fat. *Mickey Ramierz*

CROSSBRED, PUREBRED, OR REGISTERED

Your end goals may help determine whether you want to raise purebred or registered animals, or crossbred sheep. A crossbred sheep can be of two types. One is where the sire (father) and dam (mother) are of different breeds but she same breed type, such as the sire being a Cheviot and the dam being a Southdown; both are considered meat breeds. A crossbred can also mean having two parents of different breed types, such as Shropshire (meat) and Rambouillet (wool).

A purebred sheep has parents of the same breed. Pedigreed or registered animals have a known ancestry that is documented and recorded with the breed's registry association. Most sheep breeds have closed flock registries, which means only 100 percent purebred animals with registered parents can be included in the flock association. Some breeds, such as Dorper and Katahdin, allow percentage purebreds to be recorded, which is usually the result of an upgrading program.

TOP BREEDS OF REGISTERED SHEEP IN THE U.S.

- **Suffolk**
- **Dorset**
- **Hampshire**
- **Dorper**
- **Southdown**
- **Katahdin**
- **Rambouillet**
- **Columbia**

Source: 2006 USDA Agricultural Statistical Service

Crossbred sheep tend to be hardier and more productive than purebreds because of the heterosis effect, which is the increased vigor and production resulting from the crossbreeding of different breeds. Several breeds resulting from different crossbreeding combinations include the Katahdin and Polypay. Although purebred

The Cotswold is one of the oldest known breeds. It has been removed from the rare breed list of the ALBC. They make good mothers and produce quality lambs on pasture or grain. Their fleece is sought by hand spinners and fiber artists because of its quality. The rams and ewes are polled. *Bill and Liz South*

The Dorper is a hardy and adaptable breed that does well in harsh conditions and intensive farm systems. The ewes make excellent mothers and the breed crosses well with other breeds. Dorpers are non-seasonal or have an extended breeding season. They can be managed to produce three lamb crops in two years.

Hampshire sheep have the ability to efficiently convert forage into meat and fiber and are adaptable and productive in many regions of the United States. It is considered as one of the ram breeds that can produce heavy lambs. *Department of Animal Sciences, University of Wisconsin-Madison*

and registered sheep tend to sell for higher prices than crossbred and nonregistered animals, their purity or registered status does not necessarily indicate higher quality or productivity, and this may be a consideration when making your selections.

BREEDS OF SHEEP

The following is a sample listing of the many breeds available in the United States. Researching all the breeds will give you a better idea of the wide diversity that exists and help you determine which breed is best for your farm.

AMERICAN BLACKBELLY

The American Blackbelly was developed by crossbreeding programs involving primarily Mouflon and Barbados Blackbelly. They are a thrifty, small- to medium-sized sheep with a strong flocking instinct. They are productive but typically low maintenance.

BARBADOS BLACKBELLY

This breed, which originated on the island of Barbados, is more heat tolerant than wool breeds and does not require shearing. Barbados Blackbellies are very reproductively efficient because they reach puberty at an early age and have an extended breeding season.

BORDER LEICESTER

The Border Leicester originated in England from Leicester and Cheviot crosses. It is a long-wool breed medium in size and produces good mothers. Its physical characteristics consist of a white bare face with bare legs.

CHEVIOT

Originating in the hill country that borders Scotland and England, the Cheviot is a small-sized sheep with a white face and bare head and legs. This breed is noted for its hardiness and easy lambing, and is highly adaptable to a variety of climates. They can thrive on poor forage conditions.

The Karakul is considered a heritage breed and is possibly the oldest breed of domesticated sheep. It is a fat-tail sheep that has been able to exhibit true Karakul traits, such as silky pelts, even though other breeds have been introduced into its genetics. They are a hardy breed that can thrive under adverse conditions. *Department of Animal Sciences, University of Wisconsin-Madison*

COLUMBIA

The Columbia was developed in the United States from crosses between the Lincoln and Rambouillet breeds. It is a large-framed, late-maturing, dual-purpose breed that produces excellent wool.

CORRIEDALE

This breed originated in New Zealand from Lincoln, Leicester, and Merino crosses. The Corriedale is a medium-sized, white-faced breed that produces good market lambs and yields heavy medium-wool fleeces. The ewes are good mothers.

COTSWOLD

The Cotswold is a long-wool sheep with a lustrous fleece of naturally wavy curls. It originated in the hills of Gloucestershire, England, from indigenous stock and is one of the oldest breeds known. It has contributed to the ancestry of other breeds in the United Kingdom and Europe.

Katahdin sheep are named after a mountain in Maine. They are descendants of a small number of haired sheep from the Caribbean that were crossbred with the hardy breeds of the Virgin Islands and the meat and conformation of the wooled breeds. *Katahdin Hair Sheep International*

The Lincoln is a long wool sheep that is medium to large in size. Lincoln sheep have been used to develop other breeds such as Corriedale, Columbia, Bond, Polwarth, and Panama. Second-generation breeds such as Montadale and Targhee were subsequently developed in the United States from these breeds.
Brian and Jennifer Larson

DELAINE-MERINO

This is a hardy, long-lived breed that produces the finest-quality wool in the world. Delaine-Merinos possess a strong flocking instinct and extended breeding season. The breed was developed from the Spanish Merino and has an unbroken line of breeding for more than 1,200 years.

DORPER

The Dorper breed originated in South Africa from the crossing of Blackhead Persian and Dorset Horn breeds. The Dorper may have a black head or be all white. They do not require shearing and are of medium size. This breed is known for its excellent growth rate and carcass qualities. The ewes are good mothers. It is a hardy breed and adaptable to hot, dry, humid, or cold climate conditions. The Dorper is a very versatile breed.

DORSET

A dual-purpose sheep, the Dorset is medium-sized and produces a meaty carcass at a lighter weight than many other breeds. This breed is known for year-round lambing, good mothering skills, and milking abilities.

The Tunis is an attractive breed with a beautiful red color, long pendulous ears, and a calm disposition. They are very feed efficient, have a high rate of twinning, are heavy milkers, and are easy to handle due to their docile temperaments.

Polypay is a composite of four breeds that were used to develop a breed that could produce two lamb crops and one wool crop per year to give sheep producers more profits. They are medium in size and produce high quality medium wool. *Department of Animal Sciences, University of Wisconsin-Madison*

The Rambouillet is large in size and is considered one of the top ewe breeds. It is know for its long staple, dense, fine wool. They have a well-developed flocking instinct to band together in open areas. This breed is adaptable to a wide range of climate conditions.

EAST FRIESIAN

The East Friesian is the best dairy-breed sheep in the United States. The breed is of German origin and is characterized by exceptionally high milk production and long lactations.

HAMPSHIRE

Developed from crosses between the Southdown, Berkshire Knot, and Cotswold breeds, the Hampshire originated in England and is a large-framed sheep that produces well-muscled carcasses.

JACOB

The Jacob is a heritage breed listed by the American Livestock Breed Conservancy as threatened. The breed is the result of the earliest recorded breeding selection program and appears to have originated three-thousand years ago in what is now Syria. They are white with black patches and have four horns.

KARAKUL

One of the oldest sheep breeds, the Karakul originated in the deserts of Central Asia. They have long, pendulous

Scottish Blackface sheep are ideally suited to grass-based production systems. They are extremely hardy and thrifty and thrive in northern climates where other breeds may struggle to survive. *Anne and Richard Gentry*

The Shetland is a small, fine-boned breed but is very efficient in feed conversion. Their exceptionally fine, soft wool is prized by hand spinners. Rams are usually horned while the ewes are polled, although polled rams and ewes with short horns can occur. Shetlands are very easy to work with, and even though the lambs are very small at birth, they are very hardy and acclimate to a variety of climate conditions. *Cynthia Allen*

ears and a fat tail that is used as an energy reserve. This breed is hardy, adaptable, and produces easy lambers with a strong maternal instinct.

KATAHDIN

During the 1950s, the Katahdin breed was developed in Maine from crosses of the Suffolk, St. Croix, and Wiltshire Horn breeds. They do not require shearing and are easy-care sheep with a resistance to parasites.

LINCOLN

The Lincoln is a long-wool breed that is medium to large in size. Lincolns can be raised as a dual-purpose breed for their long, lustrous wool and well-muscled carcasses. Lincoln wool is favored by hand spinners.

NAVAJO-CHURRO

The Navajo-Churro is the oldest American breed and originated from sheep brought by Spanish settlers. They are a double-coated breed with a long-hair outer coat and a fine-wool inner fleece. It is excellent wool for use in hand-spinning, specialty garments, and carpets.

Southdown is considered a meat breed with an excellent carcass that yields tender, flavorful cuts. They are economical to feed and maintain, and they efficiently convert grass and smaller amounts of grain into lean red meat. *American Hampshire Sheep Association*

Columbia is considered to be the largest white-faced breed in the world. Columbia ewes make excellent mothers that have easy lambings, are heavy milkers, and have a good twinning percentage. The rams are used for range and in commercial crossbreeding programs. This breed works well in small and large farm flocks. *Nikole Riesland-Haumont*

POLYPAY

The Polypay is a medium-sized breed that produces high-quality medium wool. It was developed in the United States in the 1970s and is a four-breed composite of Finnsheep, Dorset, Rambouillet, and Targhee. Polypay ewes reach puberty early, are prolific, and have an extended breeding season.

RAMBOUILLET

The Rambouillet is a fine-wool producer that is hardy and long-lived. It was developed in France and Germany and is descended from the Spanish Merino breed. They have a strong flocking instinct and an extended breeding season. It is the largest of the fine-wool breeds and is late-maturing with a good growth rate.

ROMANOV

Originated in what was previously the Soviet Union, the Romanov is a hardy, very prolific breed that can adapt to many different climate conditions. The breed is known for early sexual maturity, out-of-season breeding, multiple births, ease of lambing, and mothering abilities. The Romanov is a double-coated breed; the sheep are born black and lighten to a soft gray as they grow their secondary fleece.

SCOTTISH BLACKFACE

The Scottish Blackface is a heritage breed that produces a coarse, long-wool fleece. They have black faces often mottled with white streaks. Scottish Blackfaces utilize rough and coarse grazing areas and are well suited to grass-based production systems.

SHETLAND

The Shetland is a small-size breed but is very hardy and can do well in rough conditions. As a pure breed, the Shetland produces a very high quality, lean meat with a flavorful, fine texture. There are eleven main whole colors in Shetland sheep, with many shades and variants including white, black or brown, reddish to fawn, and greys.

SHROPSHIRE

The Shropshire is a popular medium- to large-sized sheep that produces a quality carcass. It originated in

Targhee ewes are excellent mothers with a good milking ability. Mature Targhee ewes can raise a high percentage of twins on range conditions. They are a polled breed and can produce a heavy fleece with a typical yield between five to six pounds. The Targhee is considered a dual-purpose breed and yields a quality carcass. *Todd Taylor*

England from native stock and Southdown, Leicester, and Cotswold crosses. The breed was imported to the United States in 1855. They have a dark face and wool that extends down the legs. Shropshires possess good milking and mothering abilities and have a medium-grade wool.

SOUTHDOWN

A small- to medium-sized sheep, the Southdown matures early but produces a meaty carcass at light weights. This docile breed was imported to the United States in 1803.

The Texel breed is native to Holland and was imported into the United States in 1990. They are noted for their extreme muscling in the leg and loin and their exceptional lean-to-fat and meat-to-bone ratios. *Jan and Coby Schilder*

SUFFOLK

The Suffolk originated in England from a cross between the Southdown and Norfolk Horn. It is a large-framed sheep with a fast growth rate that produces large carcasses with a high lean-to-fat ratio.

TARGHEE

A hardy, long-lived dual-purpose breed, the Targhee has a strong flocking instinct and produces a fine to medium fleece. They are good mothers and were developed from crosses between the Rambouillet, Columbia, Lincoln, and Corriedale breeds.

TEXEL

The Texel breed was imported from Holland and is noted for its extreme loin and leg muscling with exceptional lean-to-fat and meat-to-bone ratios.

TUNIS

Tunis is a meat breed that originated in Tunisia. It was a popular breed in the South until it was almost eliminated during the Civil War. It is a medium-sized polled breed with a red or tan face and legs, pendulous ears, and no wool on their head or legs. They are good milkers and mothers and produce a medium-grade wool.

WENSLEYDALE

This is a large sheep with long-stapled, lustrous wool that falls in long ringlets in unshorn sheep. This breed was developed from a Dishley Leicester ram with a Teeswater ewe that produced a famous ram named Blue Cap, who was the founding sire. The blue pigmentation on his head and ears has become a hallmark of the breed. Rams can reach 400 pounds in weight.

Dorsets are noted for their out-of-season breeding tendencies. They are of medium size and make good mothers and milkers. They are a dual purpose breed with quality wool and well-muscled carcasses. *Department of Animal Sciences, University of Wisconsin-Madison*

HOUSING AND FACILITIES

Sheep do not need elaborate housing facilities, although housing needs may vary by climate, lambing times, and preferences. Every livestock farming operation needs facilities for storing feed, bedding, and equipment to eliminate or reduce spoilage or deterioration of machinery. The building requirements for sheep are minimal. They can survive outdoors without a barn as long as they have access to shelter from inclement weather or wind.

SHELTER INVENTORY

Sheep are social animals that need the presence of other sheep to experience a state of well-being. Solitary sheep are more easily agitated and exhibit a higher degree of stress than those involved with other sheep. Housing and facilities should provide easy access for sheep to commingle.

Important housing considerations are safety for ewes, their young, and their handler; adequate ventilation; convenience in cleaning out the bedding areas; and sufficient room for working with your flock during lambing season.

Because small flocks do not require new or expensive buildings, you may be able to convert existing buildings or sheds to fit your needs. Traditional barns, pole buildings, metal buildings, and former swine-housing units can all be refitted for sheep. Hoop buildings, which are greenhouse-type structures made of metal poles covered with heavy plastic sheeting, have recently become popular because of their low construction cost and flexible use. Metal quonset-style huts can also provide a great amount of space for a small flock and will provide protection in all kinds of weather. These huts can be easily moved from one area to another.

Buildings do not have to be large or extensive to house your flock. The main considerations for housing include an adequate amount of space for the number of sheep you have and a location in a well-drained area with easy access for manure removal.

A well-ventilated building with adequate air flow during all types of weather will provide a more natural and healthy environment for your flock. Air flow is a key ingredient to minimizing respiratory problems, particularly during times of group housing, lambing, and inclement weather.

This former horse barn has been converted into a sheep loose housing area. It has room for a large number of sheep during shearing and lambing seasons.

Whatever type of structure you have or choose to build, it should be in a well-drained area and with easy access for feeding and manure handling. If you are constructing a shed, the open side should face south and away from the prevailing winds. The best shelters are dry and well ventilated. Sheep generally will not use a shelter except to get out of very inclement weather or to seek shade. If they are given access to a shelter, they will decide whether to be inside or outside. Putting them into a close confinement area can be stressful for sheep because of the high temperatures and high humidity in the barn. Other housing options include calf hutches,

Open-side buildings with a southern exposure are ideal housing for sheep in northern climates. They provide adequate shelter during times of inclement weather and allow outdoor access for exercise. *Department of Animal Sciences, University of Wisconsin-Madison*

A metal quonset-style hut can provide a great amount of space and shelter. They can be easily moved to different fields or different parts of the same pasture if needed.

carports, and polydome structures that offer simple run-in areas to escape extreme weather conditions.

VENTILATION

Sunlight and fresh air are important for raising healthy sheep. Providing adequate ventilation in your buildings will reduce or eliminate many problems. Normal, healthy sheep can withstand most kinds of climate changes if there is access to shelter and fresh air. Respiratory diseases often arise from poor ventilation.

Depending on the type of enclosure available for your flock, you may or may not need to make alterations for air movement. If you can smell ammonia in the building, it does not have adequate ventilation.

Buildings with small entrances and exits and few windows may need fans built into the walls to provide air exchange. If the building has one open side, no fans will be required because the air will move with the natural currents. Closely confined areas, particularly during extremely hot and humid weather, will need

A hoop house is an easily erected and effective shelter for sheep. The metal frame supports heavy plastic sheeting that allows sunlight in and provides a barrier against wind and rain.

air movement to increase the comfort level for your sheep. It is better to overventilate a building than to underventilate it.

Old, infrequently used buildings with antiquated wiring can pose several safety hazards. If you are installing new fans or rejuvenating old ones, be sure that the circuits can handle the electrical requirements. Old wiring can be a safety hazard if it is inadequate to handle the electric load. Have a qualified electrician examine the wiring in your building if there are questions about the safety of old fans.

Newborn lambs are extremely susceptible to hypothermia when air movement increases, especially in cold, wet weather. Sheep require a dry, draft-free environment for lambing. When designing areas for lambs, eliminate drafts at floor level and ensure adequate air movement above the animals. This is especially important in a colder climate when your lambing season is in the spring.

PROVIDING ADEQUATE SPACE

Guidelines for the amount of space required by sheep at different periods of their development depends on whether they are confined to a building or have access to open areas. Sheep raised in confinement require more space for each ewe to prevent overcrowding. Ewes with lambs will require more space than ewes without lambs, rams, or feeder lambs. In open-shed areas, about eight square feet is typically needed for a bred ewe or a single ram. The space required increases to twelve square feet with her lambs. Because of their smaller size, feeder lambs need about six square feet each. You can determine total space requirements by multiplying the individual space requirements times the number of sheep you plan to house. They do not all need to be in the same building, but each building should have an adequate amount of space available for the number of sheep housed.

Sheep are vigorously alert to outside threats. Free-ranging animals will maintain visual contact with at least one other member of the flock. Sheep establish social dominance like any other species. The hierarchy tends to be determined by differences in age, sex, and possibly body weight. Having adequate space for your flock will diminish the effects of sheep dominance and reduce overcrowding during times of fear-inducing situations where they quickly congregate and flee.

CORRALS AND PENS

Consider dedicating an area to be used as a corral to sort and manage sheep. Tasks such as weighing lambs, trimming feet, docking tails, shearing, and giving injections are much easier in confined areas. Corrals can simply be large pens inside a building or constructed outdoors with wood or wire panels. Any form of containment where you can easily and safely work with your sheep should be sufficient.

One of the most important areas will be the pens for lambing. The lambing pen should be large enough to provide adequate room for the ewe to easily turn around, rise, and lie down. There should also be enough room for you to assist with the birthing if needed. Plans for constructing a corral and pens should be available from your county agricultural extension office.

LIGHTING AND HEAT

Sheep have a natural fear of light contrasts and may resist moving from bright areas to dark areas. A uniform lighting pattern in your buildings will reduce the effects of this fear. If sheep are raised entirely indoors, exposure

Small corrals can be constructed within a larger pen or pasture area to reduce the space available for sheep movement. This will allow you to capture or sort your sheep for such management practices as weighing lambs and trimming feet.

Thoughtful fence construction and well-placed gates can make penning sheep a simple task. By controlling sheep movements in a safe and easy manner, you reduce their agitation and decrease their risk of injury.

Placing gates in the corners of your pasture fences will make it easier to move your flock from one field to another. Minimizing stress on your sheep will make them easier to move.

A building with a uniform light pattern is the best way of reducing your sheep's natural fear of dark areas. Exposure to natural light and light patterns will minimize their hesitation to enter areas you direct them.

Traditional bedding materials for sheep include straw that is derived from small grains. The dried corn stalks seen here make an absorbent bedding material and are lightweight.

to a natural light rhythm is important because sheep are seasonal breeders. Decreasing the amount of daylight will initiate the onset of the reproductive cycle.

The temperature comfort zone for sheep is fairly wide and ranges from 32 to 85 degrees Fahrenheit, with exceptions given for newborn lambs. They will often require a more narrow temperature range, from 65 to 80 degrees Fahrenheit, for adequate comfort.

Using heat lamps near newborn lambs will provide additional warmth. The lambing area where the ewe and her lambs are kept should have access to electrical outlets where heat lamps can be used. This is important if the building is not a temperature-controlled environment. Ewes and lambs typically have a body temperature of about 102 degrees Fahrenheit with a variation of 1 degree.

Heat lamps can provide a supplemental heat source during cold or inclement weather. Lamps are placed over a spot where the lambs can congregate away from the ewe and allows them to stay warm. Poorly installed or old electrical wires should be replaced to prevent fires or overheated wires.

BEDDING

Raising sheep indoors or allowing them access to indoor shelters during changes in climate and precipitation will require the use of bedding materials that provide warmth and comfort. There are various materials that can be used for bedding sheep. Their use depends on availability and cost. Straw is a traditional bedding material for farm animals and comes from the dried stems of small grains such as wheat, oats, barley, or rye.

Dried corn stalks are the leftover residue from harvesting a corn crop. These stems can be baled, chopped, or made into small stacks. Corn stalks are highly absorbent and lightweight if loose. When baled together in small square bales, large round bales, or large square bales, corn stalks can be easily moved. If you wish to use corn stalks for bedding but don't have any corn fields, contact a neighbor who does.

Sawdust and wood shavings are highly absorbent but are not good bedding materials for wooled sheep because it gets into the fleece. It is, however, a good material to use with hair sheep. Shredded paper or newsprint is another absorbent material that can be used but is more difficult to handle. It may fly into neighboring fields or yards when it is spread on a field during high wind conditions.

Plan to purchase bedding materials before you bring sheep onto your farm. This will provide a comfortable bedding area as your sheep acclimate themselves to their new surroundings.

EQUIPMENT

Unless you are growing crops on your farm, the machinery and equipment required to raise sheep is minimal. Raising sheep, particularly if you plan to pasture them during most of the year, does not require large volumes of stored feeds or hay during winter months. In some regions the climate is moderate enough that a winter feed

Small wood huts can be used for a ewe and her lambs. These portable huts can be easily placed and moved to provide a simple solution to a shelter that may be needed quickly.

supply is available and only needs to be supplemented with limited stored feed. In most northern areas of the country, a winter feed supply usually comes from stored feed harvested during the growing season or purchased from another source.

If you have sufficient acreage for raising crops, you must decide if you want to do most of the harvesting yourself or hire someone to do the work. If you decide to harvest your crops, you will have to determine what kind of equipment you need to efficiently complete a harvest of hay, grass, corn, or other feed sources.

If you plan to raise only hay and purchase your grain needs, the minimum equipment needed will be a tractor, a mower, a baler, and a means to transport the hay from the field to your storage area, which is often a wagon. A manure spreader and a means to load the manure on the spreader, such as a skid-steer loader, will make manure hauling easier.

If you decide to do the fieldwork yourself, it is very important that you, as well as family members and

You can purchase bedding materials in many shapes and variations. They can be made from wheat, rye, or oats; and can come in small or large square bales and round bales. Wood shavings are also useful and are typically purchased in sealed bags or bulk quantities.

employees, have sufficient experience in safely and efficiently operating your equipment. Farm machinery can be dangerous to use if you don't have the knowledge or ability required to handle it properly. If you cannot find safety classes in your area or you don't feel you can do the work, one alternative is to contract others to do the field or crop work for you. This dramatically reduces the amount of equipment you need to purchase, as well as your exposure to heavy equipment.

Remember that your equipment will sit idle for the majority of the year. If you buy the equipment, you will be paying for an asset that is not working part of the time. Contract hiring of your field or harvesting work can help you get established without making a huge investment in machinery. Equipment expenditures can be substantial when getting started, and the less money spent on machinery the more that will be available for running your business. Your county agricultural extension office can help you locate the names of independent or custom operators in your area and provide typical custom rates.

Hiring someone else to do the fieldwork not only frees up capital but also relieves you of equipment repair costs. Custom operators can generally accomplish more fieldwork in a shorter amount of time than you could on you own. It is always a good idea to talk over the terms of a contract with the custom operator and develop a plan for the timely harvesting of your crops during the year well before the growing season begins.

If your farm is not large enough for custom field-work, renting equipment may be a more attractive option. Renting allows you to use appropriate equipment for the job and return it without outright purchase. Check with your local farm equipment dealers for information about renting equipment.

Many small items, including water tanks, hay feeders, forks, shovels, pails, wire, tools, gates, and numerous other items are necessary to run a farm. An easy way to learn what may be needed on your farm is to attend a local farm auction. Smaller items are typically sold at the beginning of an auction prior to the sale of large equipment such as machinery and tractors. You may not need all of the items you see being sold, but it doesn't cost you anything to attend the auction and get ideas.

Depending on the size of your flock or the number of sheep you raise indoors, manure removal will be a major consideration. The more sheep you have, the more manure will be produced. As sheep feces become mixed with bedding, a means of removing it from buildings is needed.

An adjustable chute makes loading and unloading sheep an easy and safe task. The handles allow two people to quickly move the chute from one area to another as needed.

CHAPTER 6

FENCES AND ENCLOSURES

F ences and enclosures are one of the most important considerations when raising sheep. They mark the property lines and are used to humanely enclose your animals. For generations, fences have marked boundaries and given the owners containment for their livestock and protection from animals on adjoining farms. The main purpose of your perimeter fences is to keep your animals in and your neighbor's animals out. They can also be used as barriers against predators.

If you already own a farm, it is likely you know the property lines or perimeter of your land. If you are considering purchasing a farm, there are several things to consider so that you fully understand where the property lines are located.

PERIMETER LINES

Ask to walk and view the perimeter or fence lines with the owner or real estate agent before you sign an agreement to purchase a farm: this will allow you to fully understand the boundaries of the property. If the agent doesn't know or if the party you are purchasing the farm from can't tell you, it may be worth the expense to have a surveyor establish all boundary lines around the farm before you agree to the purchase. All farms have deeds that contain legal land descriptions identifying the exact boundaries, and a qualified surveyor can use these to determine all property lines.

If the lines seem to be well established and the line fences are intact, you may feel reasonably satisfied that the land you are purchasing is what's described on the deed. If a portion of the farm has been sold in recent years, you may want assurance that the new lines are identified in the deed and that they are the boundaries for the property you are buying.

Walking the fence rows before agreeing to a purchase will give you a valuable perspective of the farm beyond looking at plat or survey maps and photographs

Perimeter fences typically serve two purposes: as a boundary line for the farm property and as a physical barrier to keep animals in or out. It is important when buying a farm to locate and fully understand the boundary lines belonging to it.

Temporary fences are very good for short-term use and can be quickly and easily moved from one area to another. They are commonly used in grazing programs.

of the property. Walking the fences will also give you a sense of proportion and provide you an opportunity to inspect the fences and make note of problem areas that need attention.

Many old fences have been removed in recent decades to accommodate larger field equipment or to allow the fence lines to be tilled to keep weeds, small shrubs, and trees from growing. If fence lines are missing, it is important to know where they belong before constructing a fence of your own. Tearing down an improperly placed fence can be costly, time-consuming, and frustrating. It may be worth the expense to hire a surveyor to identify the exact property line to ensure a fence line is properly sited before you build a perimeter fence.

Another aid to help you understand the dimensions of your property include aerial maps that have been developed by your county Farm Service Agency (FSA). Most, if not all, farming areas have aerial maps available. They provide an overview of your farm not obtained from ground level. With their assistance, the relative boundaries of the fence lines can be seen and will give some indication of the lay of the land, which may be useful in handling any hills and valleys that the fence line

Fences around the perimeter of the farm are generally made as permanent structures. They need to be in good condition in order to keep your sheep within its perimeter. A well-constructed fence will require minimal maintenance over its lifetime.

Temporary fencing can consist of simple structures such as a line wire tightly extended across the length of a field. This eliminates the use of numerous posts and is easy to erect and dismantle. Two electrified wires are used to restrain both lambs and ewes. *Department of Animal Sciences, University of Wisconsin-Madison*

traverses. You can request to see these maps prior to a farm purchase by visiting your county's FSA office, and you can obtain copies once you have bought your farm.

One of the benefits of these maps is that they are developed in conjunction with farm programs and provide excellent guidance to determine which areas of the farm are best suited for different purposes. With the FSA's help, you will be able to identify areas suited for hay or permanent pasture, woodlands that may or may not be pastured, and areas suitable for cultivated crop production. These land assignments are made based on uses that will best protect the soil structure and return the greatest possible profit for each type of soil.

PLANNING YOUR FENCE

Raising sheep will typically require only two kinds of fences: permanent and temporary. Permanent boundary or division fences require different materials than fences for temporary lots. Permanent fences are intended to last many years with minimal repairs and should be constructed with sturdy and high-quality materials. These are typically perimeter fences or fences around streams or waterways. Temporary fences, usually found inside perimeter fences, are intended to subdivide fields into smaller areas called paddocks and are constructed to last

Polytapes, polywire, and metal wires are all flexible enough to be unrolled for use and rolled back up for storage. Storing them in a dry area will help lengthen their lifetime and usefulness.

only a short time or can be easily moved from one area to another.

If the perimeter fences are in good repair and appear to be free of holes or downed wires, you may need to do little construction work. However, if holes or breaks in the fence line exist, it is wise to fix them to avoid escapes by your sheep or an easy entrance for predators.

Woven wire is a good fencing material to use with sheep. A woven-wire fence, when correctly built, will maintain its strength and effectiveness for many years.

If you need to construct a new fence, the key to successfully building one starts with a good plan on paper because alterations can be made quickly if you change parts of it. Drafting a fence on paper allows you to calculate the length of the fence and the amount of fencing materials needed before you drive in a post. It will also help you estimate the cost of your fencing project.

After you have developed your plan on paper, take a walk out to your fields to see if your ideas will work or if you need to make changes before you start. It is much easier to redraw your plan on paper than to tear down a partially constructed fence and start over.

Certain areas of your farm may be more suited for pastures than crops but may be located a long distance from your buildings. In this case, you may need to construct one or more lanes or pathways for your sheep to gain access to those areas. A key rule is to build straight fence lines wherever possible. They are easier to construct, retain their tension for a longer period of time, and require fewer materials. You may need to make curved fences in certain areas, but try to avoid these whenever possible.

It will probably be less expensive for you to construct your own fences rather than hiring out the work. You can hire fence contractors who will finish the work

The spacing between the horizontal wires increases with fence height. The spacing at the bottom of the wire should be no more than six inches for most sheep breeds. For some of the smaller sized breeds, such as Shetland, the spacing at the bottom of the wire should be no more than four inches.

Small spacing between the horizontal line wires of woven wire fencing will reduce the risk of a sheep pushing its head through the wires and getting caught. Daily observation of your flock will alert you to any problems that may arise with sheep getting caught in the fence.

Wood posts make great stabilizing anchors for fences. They are particularly useful when placed on hilly terrain or at corners. They can be used exclusively in a fence line or spaced between steel T-posts to help strengthen the fence.

quickly (at a greater cost), but with a little experience and guidance you can build a fence that will withstand the pressures placed on them by your sheep or weather conditions and will last a long time.

PERMANENT AND TEMPORARY FENCES

Good fencing protects and confines valuable animals by providing a barrier to restrict their movements. A permanent fence that surrounds your farm is one of the best ways to protect your livestock investment. Besides establishing a fixed property line, perimeter fences are the last line of defense if your sheep happen to escape from their designated pastures, feeding areas, or small lots around your buildings. Perimeter fences will also keep your sheep from wandering onto highways and roads and will protect them and the driving public from possible highway collisions. If your perimeter fences are intact it is unlikely your sheep will be able to invade your neighbor's property, thereby relieving you of possible financial liabilities due to destruction of property or crops. You cannot decide what species or stocking levels your neighbors may have, so maintaining good fences for your farm will protect your property by keeping other animals out.

Consider a permanent fence for those areas used for pasture several years in succession. Farm ponds and other waterways should receive priority for permanent fences to control access or allow access only for drinking. You may also consider permanent fencing for fields where cultivated crops are grown and your sheep are allowed grazing access after harvest.

If it is not possible to construct a permanent fence around the entire perimeter of your farm, consider building those sections that will be most useful to your enterprise and plan to add the rest at a later date or by incremental construction of the rest.

Temporary fences are used for a few days, weeks, or months and can be removed. Movable fences are less expensive to build than permanent fences and are quicker to set up and take down. They are less durable and may not be as effective, particularly if maintained by an electric current.

Temporary fencing has several advantages in being lightweight and easy to move. Temporary fences offer flexibility for pasturing sheep by expanding or contracting the size of the paddock or pasture. This flexibility allows you to accommodate any increase or decrease in the number of sheep placed within a pasture area as well

Steel T-posts are commonly used for fencing. They are driven into the ground with a post hammer. Plastic insulators can be attached to T-posts at varying intervals to hold horizontal wires. Wire tighteners can be used to keep them taut.

Wire paneling can be used in high traffic areas or to corral sheep. The smooth wires of the panels will not injure sheep or their wool. They are sturdy, easy to erect, and last many years.

as their relocation from pasture to pasture, and provides ease in developing different pasture rotation schemes.

FENCING MATERIALS AND VARIETIES

Materials for permanent fencing most appropriate for sheep include woven wire, high-tensile wire, mesh wire, wood boards, or wire panels. Barbed wire is typically not a good choice because it does not effectively deter predators and can injure sheep if they happen to get caught between several strands. Wool snagged in the barbs may affect its quality. Barbed wire can be used at the top of a woven wire fence as a deterrent for other animals or predators. One or two strands will generally be sufficient. One strand can be laid along the ground at the bottom of the woven wire as an additional deterrent.

Temporary fences are typically made from materials that can be easily moved, such as a single metal wire, poly-wire, or wire tapes. These lightweight materials can be rolled up on a spool and moved from one location to another with minimal effort and can be quickly unrolled to create a new temporary fence.

Your budget may influence the types of fencing materials you decide to use. Although there are differences in the cost of fencing materials available, it is usually wise to consider purchasing materials that will last a long time. If your plans and goals change during the course of your farming career, you can use your well-constructed fences for a variety of livestock without making major alterations or additions. While the cost of fencing materials is a major factor in your decision, you should consider balancing that cost with the performance and time element involved. The initial cost of constructing a fence is spread over the life of the fence. A well-constructed fence that requires minimal maintenance and repair over its lifetime is an excellent investment in the protection of your farm and sheep. If you are planning to construct fencing around your entire property or fractions of it, you can consider using a combination of fencing types to suit different areas and possibly reduce the cost.

WOVEN-WIRE FENCES

Woven-wire fences consist of a number of horizontal lines of smooth wire held apart by vertical wires called stays. The height of most woven-wire fencing materials ranges from twenty-six to forty-eight inches. The spacing between horizontal line wires may vary from one and a half inches at the bottom for small animals to eight to nine inches at the top for large animals. Wire spacing typically increases with fence height. The spacing between the stays can also vary, and you should consider using woven wire with stay spacing of no more than six inches for sheep or other small animals. You can allow for twelve inches for large animals. The manufacturer's label attached to the wire roll will tell you the type and style of the roll by using standard design numbers and will help ensure you are buying the correct fencing product for your needs.

Although sheep are agile, they are not athletic enough to jump high fences; you may want to consider using short wire rolls rather than tall rolls to lower your costs. Woven wire is often more expensive than other types of fencing materials because of the additional metal used in the design. Woven wire can be two to three times the cost of barbed wire for a similar distance covered while having approximately the same life span. Woven wire is sold in rolls, which are heavier and more difficult to handle, and more roll units are needed to complete a similar distance when compared to barbed wire. Even if you are faced with these challenges, woven-wire fences make the best structures for sheep lots, corrals, permanent pastures, and areas where close confinement is needed.

HIGH-TENSILE FENCES

High-tensile electric fences consist of a single smooth wire that is held in tension between end-post assemblies. They are relatively easy to install, last for a long time, cost less than other types of fencing, and are easily adapted to specific needs. When used with sheep, multiple strands—as many as five or six—are needed to keep them in and predators out because the bottom strands are more closely spaced than the top wires. High-tensile wires are held on wood fence posts with plastic insulators or can be threaded through plastic posts or metal posts using insulators. High-tensile fences require strong corners and end braces to achieve adequate tension.

It is important to keep the fence electrified to maintain the integrity of the wire. High-tensile wire can stretch if rubbed or pushed against by animals. This stretching will cause the wire to sag and leave gaps for the sheep to crawl through.

One of the disadvantages of high-tensile fencing, particularly in climates that receive large amounts of snowfall during the year, is that snow acts as a grounding agent and the electric current's effectiveness is diminished as it becomes buried under large snowdrifts. Although your sheep might not challenge a high-tensile fence buried in snow, the stretching effect on the wires as the snow melts will cause the line to sag. High-tensile wires are typically attached to a clamp-and-spring assembly at the corner posts so that the wires can be sufficiently tightened in the spring.

An electric fence charger or controller is used to transfer and convert the flow of the electric current in your outlet to a series of short pulses that travel along the pasture wire. Controllers need to be correctly grounded to be effective in pasture use.

ELECTRIC FENCES

Electric fences offer the flexibility of being used for both permanent and temporary fencing needs. They are psychological barriers more than physical barriers: the only thing that would keep your sheep within a designated area by using a single wire stretched across a field is the shock the sheep receive when they touch the electric wire. The purpose of the shock is to scare or surprise the sheep when they touch the wire. It alters their behavior by training them to avoid the wire in the future and is an effective and humane management tool for sheep.

Used as a movable or temporary fence, electric fences can be made with one, two, three, or more strands of smooth wire or a poly tape that has small wires woven into it. The poly tape is more flexible and easier to handle and move from one location to another than solid wire. Metal wires, poly tape, or high-tensile fences that are electrified are typically energized by an electric controller that receives its source from a standard farm electrical outlet or a solar-powered pack. Both fence types need to be grounded to complete the circuit. Poor grounding is one of the leading causes of electric-fence failures. Your electric fence must be properly grounded so that the pulse can complete its circuit and give the sheep an effective shock. Always follow the manufacturer's instructions for grounding electric fences.

Electric fences can be useful in conjunction with permanent fences to provide a second barrier of containment. This is important if you have a permanent fence that may not be in the best condition and you do not intend to replace it. You can accomplish the same result by positioning the wires near the permanent fence.

The electricity used in the wire is generally harmless to animals or humans, although it can cause momentary discomfort. A controller or charger regulates the flow of energy by supplying short pulses of electricity that travel along the wire. When a sheep comes in contact with the wire, it completes the circuit from the wire through its body and to the ground. This sudden shock or jolt will discourage further contact with the fence. Sheep require training when first using electric fences. They will often not be aware of the fence until encountering a shock from the wire. It's important that the fence charger be maintained at full-time operation for the temporary electric wire to be effective in keeping sheep inside the fenced area.

An electric fence is typically not continually electrified but remains effective because of short electric bursts. In many states it is unlawful to use any electric fence unless a controlling device regulates the charge on the fence wire.

Solar power packs rely on the same principle as electric control devices but have the advantage of not using an electric source for which you pay. Solar packs derive their energy from the sun and deliver the current through the wire. Once some packs are fully charged, they have the capability of providing a low impedance current without the sun for up to two weeks. These may be useful in areas that are a far distance from your buildings. Battery-powered charges are also an option, but they tend to have a shorter lifespan before they have to be recharged or replaced.

The effectiveness of an electric fence diminishes when the electric current no longer runs the length of the wire or when vegetation grounds the wire. To keep your electric fence in top condition, periodically check along the length of the fence to make sure the current is traveling through the entire fence. Keep the grass underneath the fence trimmed so that no vegetation touches the fence to ground the current.

Do not use homemade or inexpensive high impedance controllers. These devices may cause serious injury or death to sheep or humans. The use of poorly designed controllers may result in grass fires along the fences. Under no circumstances should your electric fence be directly connected to a live electrical outlet or high-voltage electric source.

WOOD FENCES AND WIRE PANELS

Using wood boards for a fence is an option that provides a strong barrier and an attractive feature to the farm. Board fences are safe for animals and those working with them because they lack sharp points that may cause injury. One disadvantage of board fences is the higher cost to build and maintain them. It also requires more labor to plan and build a wood fence than a wire one, although posts need to be set in either system. Board fences need to be attached to wood posts, but woven-wire, electric, high-tensile, and poly-wire fencing can be attached to steel or wood posts depending on the purpose and preference of the farmer.

Board fences are usually made from one- to two-inch-thick, four- to six-inch-wide, eight-foot-long boards. A wood-board fence typically will be four and a half to five feet high. By spacing the boards four to six inches apart (depending on your preference), you can calculate the amount of materials needed for your sheep lot or pens. Because of the higher cost of board fencing, it is generally used in corrals or outside pen areas instead of the whole farm.

While wood fences may be useful in small areas and around buildings, they also have a shorter lifespan because of weathering effects and often begin to splinter, break, or rot after a number of years unless periodically repainted to repel the weathering effects.

Wire panels are an alternative to wood fences and, in some cases, to different electric fence types. Although

Above: Wire netting can be used in gateways or lanes. It has the advantage of providing an electric barrier and is easy to step over or open without touching the electrified threads.

Left: Breakers can be installed at different points along your electric fence. This allows you to engage or disengage the circuit as needed. Planning the layout of your electrical fence will save you time and materials and create an easily used, effective system.

Lanes can be constructed to allow easy transfer and movement of your flock from one field to another. One major lane system built to service all of your pastures can include a gate system that opens into individual fields or paddocks. Plan the layout of your fencing system before you build it so it will accommodate ease of movement and have the most effective use of materials.

they are similar in construction to woven wire, these panels are made of heavier metal material and are welded at the joints to provide a sturdy, long-lasting fencing option. Wire panels can also be used in close-confinement areas, such as lots, small pastures, loading areas, and around buildings.

Wire panels are sometimes used as sturdy alleyways between buildings and can be easily erected and quickly taken down. There are many and varied uses of wire panels that provide the flexibility of a fence without it needing to be either permanent or temporary.

ADDITIONAL MATERIALS

If you are constructing a fence, the most important item you will need is a good pair of leather or heavy cloth gloves. Such gloves are absolutely essential when working with wire as they reduce the risk of injury and severe cuts to your hands and fingers. Some of the tools needed to complete a fencing project include fencing pliers, a posthole digger, protective eyewear, a tape measure or reel, and a wire puller. When you have determined the type of wire that will work best for your situation and budget, you will also need staples, wood posts or steel

T-posts, and clips. Many of these items can be purchased from farm supply stores, lumberyards, farm catalogs, hardware stores, or Internet companies. It is usually less expensive to find a local source when purchasing a large quantity of wire fencing, boards, posts, braces, and other heavy items. Shipping costs can quickly mount from other sources unless they are included in the total price. Be certain to fully understand any shipping costs involved whether or not you purchase from a local business. Sometimes you can receive a discount on large volume purchases and delivery of these materials if you can't haul them yourself.

Different classes of zinc coating on fencing wire have been established. Generally speaking, the higher the class number, the greater the thickness of the zinc coating on the wire, which leads to a longer life. Dealers in wire fencing can offer advice on the type and gauge of wire that will best suit your needs.

Fence posts are used to hold the wires apart or to keep woven wire fences erect and secure. They are commonly made of wood or steel; both have advantages and disadvantages. Wood posts have an advantage over steel posts in strength and resistance to bending. Permanent fences often require decay-resistant fence posts. Wood posts that have been pressure treated can last as long as steel posts. Wood posts come in varying sizes, so it is important that you use larger-sized posts for the corners and braces. Smaller-sized posts can be used as line posts.

Wood posts need to be long enough to support the fence height and depth when they are placed into the ground. A satisfactory post length is the combined height of the top wire above the ground and the depth of the post in the ground, plus six inches.

Steel posts have several advantages in that they cost less than a similar wood post, can be driven into the ground more easily, and weigh less for easier handling. Steel posts generally are from five to six feet in length.

The main ingredients to building a permanent fence with a long lifespan are solid end or corner posts, tight wire, and the use of good materials. Every fencing job has different requirements and each fence presents a slightly different approach. Like other construction and maintenance jobs around the farm, building a

Sturdy gates that are easy to open and close will aid in moving your sheep. Corner posts can anchor the gate and the electric fence that can be buried at traffic areas and surface at a breaker point.

good fence requires proper techniques and a common-sense approach.

FENCE CONSTRUCTION

Clearing away brush, shrubs, and old wire will make the construction of a new perimeter fence much easier because it will leave you a clear path in which to work. If this is not possible, you should at least try to make the fence line clear of major obstacles that would impede your work.

If you are constructing a completely new fence, the first step is to locate the corner or end posts. If you are simply repairing a fence, this is not an important consideration. Locating the corner posts will allow you to

A foldable V-gate is useful when trying to corral one or two sheep. Light in weight but sturdy, these types of gates are easily moved.

plan your fence layout because you will be aiming for the corners as you unroll the wire. You may also want to determine the placement of gates if you decide to include them in your fence structure. Gates and passageways for your sheep should be located in the corners of fields nearest to the farm buildings. Placing them in corners makes it easier to move sheep from one field to another and allows you to have your corner posts and gates in areas that do not break up the fence line.

Two of the most important contributing factors to a durable, long-lasting fence are well-set corner posts and tightness of wire, and both are dependent upon each other. When wire is stretched, the pulling force on the corner post may reach 3,000 pounds. Winter cold can cause contraction of the wire, which can increase the pull to 4,500 pounds. The corner and end-post assemblies must be strong enough to withstand these forces or the posts will slowly be pulled out of the ground and the wire will lose tension.

If you need to replace corner or end posts you should set them at a depth of three and one-half to four feet. In colder climates, the shallower the depth, the more likely the post will eventually work its way out of the ground due to ground heaving in spring. A post moving vertically, even two inches, will cause the fence to slacken.

Corners should have braces positioned between the end post and a second post that is set at a distance of two and one-half times the height of the fence for maximum support.

Once you have determined the approximate fence line, you can lay down a single strand of wire from one end post to the next. When tightened, this will provide a straight line for placing your wood or steel T-posts at appropriate distances. Once the posts are set, you can unroll your woven wire and stretch it with the use of a woven-wire stretch assembly. Using this tool will keep the wire from stretching out of shape and evenly stretch the entire roll. One caution about tightening a long roll of woven wire that has been laid out: During this tightening, the wire will often have a tendency to want to turn over on itself. Make sure the wire remains flat to get an even stretch.

The spacing between wood line posts is normally twenty to twenty-five feet, and the spacing between steel posts is normally five to eight feet. It all depends on the length of your fence. If the total length of the fence is greater than 650 feet between corner posts, it is advisable to insert a braced line post every 600 to 650 feet. This will help keep the fence tight over long distances. Braced line posts are also very useful when the fence is made over rolling land or hills. Once the layout for the perimeter fence is finished, you can plan the interior fences, corrals, waterway barriers, or other enclosures.

If you are going to create lanes for your sheep, it is best to locate them in the driest areas possible, such as along a natural ridge or some other higher land feature. This will allow you options for livestock usage if your future plans change.

QUESTIONS ABOUT FENCING

Booklets explaining the different fence construction options are usually available from county agricultural extension offices or fence manufacturing companies. The initial costs of fence construction are usually high; there may be programs available through land conservation agencies to help offset some of these costs when applied to certain management practices. Check with your county agricultural extension office to learn more about available programs.

Fences can be built along windbreaks to offer protection from the elements. Windbreaks can also be planned and planted after fences are constructed.

FEEDS AND FEEDING

The daily nutritional requirements of each sheep on your farm must be taken into consideration to provide adequate growth, maintenance, and reproductive efficiency. Raising sheep during the summer is relatively easy with adequate pastures. Sheep have higher nutritional requirements during the winter, and understanding these requirements will help you formulate rations as well as plan for the crops required to meet these needs. You can lower feeding costs by incorporating a grazing program, which utilizes the ruminants' efficient way of converting forage conversion into meat or milk.

SHEEP ARE RUMINANTS

Sheep are ruminants, which means their stomachs have four chambers. These chambers are basically small fermentation vats that break down plants and grain into carbohydrates, proteins, enzymes, and other components needed for growth, maintenance, or milk production. This specialization gives sheep the ability to absorb nutrients and break down grasses through acidification. This allows ruminants to extract nutrients from low-quality feeds and efficiently utilize plant products that other animals cannot use. In this way they become a conduit for products not edible by humans and convert them into food, such as meat and milk, or wool products that humans can use.

The symbiotic relationship between the microorganisms that exist in the front part of the stomach is the key to the ruminant digestive system. These bacteria and protozoa have the capability to convert solid plant material into enzymes usable by sheep. As these microorganisms work together, they create cellulase, an

Sheep are ruminants and can make use of a wide variety of feedstuffs including grass, hay, grains, weeds, and silage. They also require minerals and vitamins. Avoid feeding any type of moldy feed to sheep because it can incubate harmful bacteria. *Cynthia Allen*

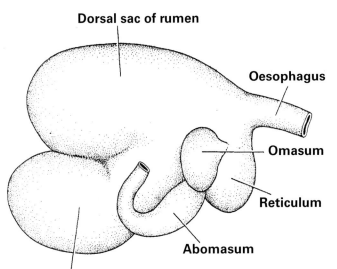

The ruminant has a stomach made up of four chambers. The rumen, reticulum, omasum, and abomasum work together to break down feedstuffs and convert them into proteins, energy, fats, and other nutrients useful for growth and maintenance. *Department of Animal Sciences, University of Wisconsin-Madison*

Dorsal sac of rumen

Oesophagus

Omasum

Reticulum

Abomasum

Ventral sac of rumen

enzyme that animals can't produce on their own. Cellulase enzymes break down the cell walls of plant materials, which releases the plant's fatty acids into the digestive system. These fatty acids are then absorbed by the ewe or lamb and make a significant contribution to its overall energy needs. As the other plant materials pass into the intestine, amino acids, lipids, carbohydrates, and other enzymes are absorbed in much the same way as other animals.

The ability to use these microorganisms to aid two different digestive systems allows sheep to use cellulose as an energy source that other plant-eating animals cannot. This makes sheep ideally suited to eat crops produced on poor soils that may be indigestible or unsuitable for other species.

SHEEP DIGESTIVE SYSTEM

The four chambers of the ruminant digestive system— the rumen, reticulum, omasum, and abomasum—work together in a specialized system. The rumen is the largest chamber where the grass eaten by the sheep first enters the digestive system. The rumen becomes filled with chewed and half-chewed materials that are mixed with saliva as the sheep regurgitates and swallows the materials several times. The chewing process grinds the food into smaller portions and injects saliva into the material as it is crushed and chewed. This crushing allows more surface area of the material to be exposed to the bacteria in the rumen.

Cost and availability of feedstuffs are two factors that sheep producers must consider to make sure the rations they formulate will be low cost yet nutritionally adequate. *Department of Animal Sciences, University of Wisconsin-Madison*

Pastures are the most economical source of nutrients for mature sheep and lambs. Grass grows readily in most regions of the country and can be a ready source of feed for your flock. *Department of Animal Sciences, University of Wisconsin-Madison*

As the fiber breaks down into cellulose, the bacteria and protozoa break down the cellulose into cellulase and then into glucose, which is used by the microorganisms to feed themselves. While some microorganisms escape the rumen and pass through the other chambers, most stay behind to work on the newly ingested plant materials.

The broken-down fiber immerses itself in the saliva and stomach acid, the liquid portions, and then passes through the rumen to the next chamber, the reticulum, and then to the omasum, where the water is removed. Throughout this process, the materials are constantly being churned to mix the liquids, solids, and bacteria to keep the fermentation process going.

As the water is being removed, the material moves along into the next chamber, the abomasum, where it becomes digested much like it would be in the human stomach. The abomasum, unlike the rumen, reticulum, and omasum, does not absorb nutrients. It prepares food for the enzymatic breakdown and absorption in the small intestine.

From the intestine, where the absorption of nutrients continues, the unusable portion is finally expelled as feces. The digestive process of ruminants produces acetic acids, propionic acids, and butyric acids, which are the volatile fatty acids produced by the bacteria that give the ruminants their energy.

Lambs do not have a functioning rumen or reticulum when they are born and colostrum milk is vital to their survival. As they begin to ingest dry feeds over the next few weeks, whether in the form of a supplemental feed known as a creep feed or hay, the microbes begin to multiply and stimulate the development of the rumen and reticulum, which then are usually functional by the age of six to eight weeks.

FORMULATING CREEP FEED:
- Creep feed should contain 18 to 20 percent crude protein because the growth of young lambs is mostly lean muscle rather than fat.
- The creep feed should be all natural grains and proteins—no urea.
- It should contain a 2:1 calcium to phosphorus ration to prevent urinary calculi (kidney stones) in male lambs.
- The creep feed should contain a coccidiostat to prevent coccidiosis.
- Lambs should be vaccinated with clostridium perfringins C and D to prevent overeating disease prior to weaning.

GRASS FOR YOUR SHEEP

Most of the nutrients sheep need for growth and production can be obtained from good-quality grasses alone. Grasses, clovers, and alfalfa are the most abundant natural resources available for feeding sheep and are the least expensive crops to produce and harvest. Grasses grow plentifully in many areas of the country and can be used as a major source for feed and roughage.

Grasses can grow virtually everywhere, particularly in the Midwest and upper tier of the country. There are a number of varieties available to provide the nutrients necessary for an adequate rate of gain. Understanding the basics of plant growth is essential to establishing and maintaining pastures that adequately support your sheep for the full growing season.

PLANT GROWTH

The key to good pastures is a fertile soil that takes advantage of photosynthesis, which transforms sunlight into the energy plants need for growth. This energy is converted into carbohydrates, which can be used for growth or stored for future use. Because photosynthesis occurs in the leaves of plants, the plants grow slowly at first due to the small surface area of the leaves. This is evident in the early spring or after a cutting of grass or hay has been removed or grazed. Referred to as recovery time, the plants are using some of the stored energy in their roots during this phase to start growing again. As the leaves enlarge, photosynthesis dramatically increases and the plants grow at a rapid pace. As the plant matures, its growth slows as it develops flowers to produce seeds. At this point, just before the flowering stage, the nutrient quality of the plant material reaches its peak.

The quality of the plant dramatically decreases and its nutrient content diminishes as it matures. A greater percentage of the plant's nutrients become tied up in nondigestible forms, such as lignin, as the plant ages. As

The nutrient requirements of sheep depend upon their size, which is largely due to breed characteristics. The Wensleydale on the left towers over its Shetland companion and requires a larger amount of daily nutrition. Consider the current body condition of your sheep versus what is desired to formulate your feeding program. *Cynthia Allen*

As their lactation progresses, ewes produce less milk per day and require less grain to maintain that production. As the fall breeding season approaches, ewes will prepare for pregnancy rather than producing milk.

than a stand of only one variety. Having multiple varieties growing in a pasture allows more flexibility in climate conditions occurring during the year. It also provides more stability in changing weather patterns because each species has different strengths.

Some grasses, such as timothy, clover, and bluegrass, prefer cooler temperatures and are more productive in the spring and fall. Legumes, such as alfalfa, start growing a little later in the spring but have a uniform growing pattern during the summer season. There are grasses

Lambs quickly learn to follow their mothers and the rest of the flock as they graze on pasture. Although they initially rely on milk, lambs will utilize forages in their diets within a matter of weeks.

the amount of nondigestible fiber in the plant increases, the result is lower-quality forage with a decrease in total digestible nutrients (TDN). TDN is used in feed ration calculations to help determine the contribution of the forage to sheep dietary requirements.

If you are using a rotational grazing program for your flock, you can circumvent the problem of fast-maturing, low-quality forage by moving the sheep through the rotation more quickly so that they eat the tops of the plants rather than the full plant. This will slow the accelerating plant growth and allow you to harvest a crop for winter feed while it still retains a high level of nutrients.

TYPES OF PLANTS

There are many varieties of grasses and legumes that can be grown to supply the nutrients your flock needs. You may want to consider a mixture of several grasses rather

Creep feeding is advantageous for flocks that have many multiple births or in flocks where milk production is a limiting factor. Lambs will begin to eat solid feed between one and two weeks of age, although they will not eat significant amounts of grain until they are three to four weeks of age. Most lambs can be successfully weaned at sixty days of age or forty-five pounds, whichever comes first. *Mickey Ramirez*

that thrive on the midsummer heat, which normally slows down many other types. These are warm-season grasses and include bluestem and switch grass, which tolerate very dry conditions.

Oats, wheat, and winter rye are grasses that have very good quality in the spring, but their stems are generally unpalatable and better used for bedding as they mature. While their seeds are good in rations, their quality drops dramatically as they mature.

The species of plants used in your pastures should be tailored to the grazing system you have designed for your region. The climate, soil type of your farm, amount of moisture your area typically receives each year, length of your growing season, and number of sheep you want to put on your pastures are all considerations for the kind of grass and legume species that should be seeded into your pastures. A list of grasses and hay varieties best suited for your area can usually be obtained from your county agricultural extension office.

A starter ration (19 percent crude protein) can be provided to lambs at an early age:

Ration	Percentage
Rolled Shelled Corn	47.8
Rolled Oats	12.5
Premix with Bovatec	16.5
Soybean Meal	17.2
Molasses	5.0
Sheep Mineral	0.5
Ammonium Chloride	0.5
	100.0

NUTRITION REQUIREMENTS

There are three stages you will need to consider in your sheep-raising program. They are growth, gestation/lactation, and maintenance. Each phase has different nutritional requirements. Those requirements are greater

Creep feeding is a means of supplying extra nutrition to nursing lambs. This is usually in grain form. Creep feed helps teach lambs how to eat dry feed and stimulates the development of their rumens, which aids early weaning. A feeder like this can be used for creep feeding.

Milk replacer can be dispensed to a large group of lambs by a machine. Long plastic tubes extend from the machine to the pens and carry milk to nipples that lambs can nurse on. Routine cleaning of the machine and milk tubes is essential to minimize the risk of harmful bacteria. *Department of Animal Sciences, University of Wisconsin-Madison*

during the growth stage as the feed and forage ration is converted into muscle and bone development and daily maintenance. The lactation stage requires nutrients to replace losses in the ewe's body composition and provide adequate elements in the milk for the lamb. Maintenance is the replacement of nutrients lost through daily movements and body functions.

The diet you devise must provide all the essential nutrients, vitamins, and minerals for proper health and growth. It will be determined, in large part, by the availability of feedstuffs and their costs, the feed processing and handling equipment available, and the productive phase of the sheep. Failure to replace nutrients lost through these processes will result in poor performance, poor health, and loss of body mass.

Guidelines for the daily nutritional requirements of sheep can be used to formulate typical rations. While individual animals may differ in their requirements and growth rates, a ration that meets the needs of the majority of the flock will generally suffice. Those sheep that have accelerated metabolism levels tend to eat more and can be accommodated with more forage. Another factor that affects your ration program is the amount eaten daily by an individual sheep. This will vary with the weather conditions, health of the animal, feed palatability, age, and size.

The nutritional requirements for ewes increase during late gestation and lactation versus the maintenance and early-gestation periods. For a typical 150-pound ewe, you should provide daily nutrient requirements in 3.5 to 4.0 pounds of feed. Of this amount, the overall fiber content consumed should be a minimum of 40 percent, or about 1.5 pounds, which leaves 2.0 to 2.5 pounds for grain.

The daily grain rates for maintenance and early-gestation ewes should be between 1 to 2 pounds as a supplement to hay or silage. They should generally be allowed full access to hay. The forage supplement may be as much as 2 to 4 pounds of hay or 7 to 9 pounds of corn or hay silage. For late-gestation and lactating ewes, 3 to 5 pounds of grain can be provided with 4 to 7 pounds of hay and 10 to 20 pounds of corn or hay silage. Feeder lambs can be fed 2 to 3 pounds of grain and 4 to 6 pounds of silage, plus protein and mineral supplements.

Late gestation and early lactation are two stages when the required amount of nutrients must be supplied to avoid problems later on. Anticipating these changes pays dividends in healthier ewes and lambs. *Gary Jennings*

There are differences in feeding a farm flock when compared to range ewes. Range ewes typically only need to be fed if they are in inclement weather or if they are confined for shed lambing. They may also be supplemented during breeding, late gestation, early lactation, or during drought periods when forage is not available. They can be fed the same rations as the farm flock.

DAILY REQUIREMENTS OF SHEEP (PER ANIMAL)
Grain
Ewes—Maintenance and early gestation: 1 to 2 pounds as a supplement to a hay or silage ration

Ewes—Late gestation and lactation: 3 to 5 pounds as a supplement to a hay or silage ration

Feeder lambs—2 to 3 pounds plus mineral supplement

Hay
Ewes—Maintenance and early gestation: 2 to 4 pounds

Late gestation and lactation: 4 to 7 pounds plus supplemental grain

Corn or Hay Silage
Ewes—Maintenance and early gestation: 7 to 9 pounds

Late gestation and lactation: 12 to 20 pounds

Feeder lambs—4 to 6 pounds plus 1 to 2 pounds a day along with protein and mineral supplement

Water
Ewes—Free choice

Feeder lambs—2 gallons daily or free choice

FEEDSTUFFS AND SUPPLEMENTS
While pastures, grasses, and hay are the most natural nutrient sources for your sheep, other feedstuffs can provide for their requirements. You can substitute one for another as long as their nutritional requirements are being met and nutritional deficiencies or imbalances are avoided.

Pastures are the most economical source of nutrients for sheep and lambs and may provide all their needs.

The most profit is usually made by feeding sheep for optimum production. Some smaller flocks can utilize low-cost, low-energy feeds to prolong the feeding period to try and capture a better price during seasonal market fluctuations. Sheep can forage on field waste products such as corn fodder (foreground) and turnips (background). *Department of Animal Sciences, University of Wisconsin-Madison*

Sheep are excellent weed eaters and may choose to eat weeds over grass. They can be used to control invasive or obnoxious weeds. Hay is usually the primary source of nutrition during the winter months and can vary in quality because of maturity or the conditions under which it

is harvested. Knowing the nutritive value of your forages will be helpful in formulating your ration to meet those requirements. Check with your local feed mill or county extension agricultural agent for companies that can provide a certified forage test for your use.

Silage, whether it is corn or hay, is a good feed for sheep and refers to high-moisture forages—usually corn, hay, or oats—resulting from fermentation. Because of its bulk, sheep often cannot consume enough silage to meet their nutritional needs. Although silage is typically fed on larger sheep farms or ranches, small- to medium-sized flocks can utilize silage, which can be easily stored in plastic silage bags. These airtight environments promote proper fermentation and the feed can be accessed as needed.

Regardless of whether the silage is stored in upright silos, sealed plastic bags, or some other form that facilitates fermentation, it is extremely important to feed your flock good-quality silage. Avoid using silage that is moldy. Feeding your flock moldy silage, hay, or grain can be deadly. Mold can incubate harmful bacteria that can cause listeriosis, a condition that causes "circling disease."

It is often necessary to feed your flock grain or concentrates to provide the nutrients that silage or pas-

Yearlings with lambs need a higher level of nutrients than mature ewes in order to accommodate lactation requirements and growth. Supplying inadequate nutrients during this stage will lower conception rates and weight gains. *Department of Animal Sciences, University of Wisconsin-Madison*

Above: Wheat and oats are cereal grains that can be grown on most farms and used as part of the ration mix. They are high in total digestible nutrients (TDN) but lower in protein value.

Left: Clovers are an excellent feed source and are high in protein and nutrients. When used in a rotational grazing program, grasses and legumes can provide a high percentage of the diet and help reduce production costs.

The daily nutrient requirements of a typical 150-pound ewe should be contained in about four pounds of grain mix. This can be supplemented with pasture grasses or baled hay.

There are many different types of feeders available that can be used indoors or outdoors. This style has an overhead rack to hold loose hay with a lower pan for grain. Loose leaf particles from hay that are high in protein are saved when they drop into the pan.

ture grasses cannot, particularly with lactating ewes. Although high in energy and protein, pastures may also be high in moisture content. This combination makes it difficult for high-producing sheep to eat enough grass to meet their nutritional requirements, and they must have a grain supplement.

Grains or concentrates provide energy and protein. Energy feeds such as corn, wheat, barley, milo, and oats are usually high in TDN but tend to be low in protein. One concern with feeding high levels of cereal grains is that they tend to be high in phosphorus and low in calcium. High-phosphorus, low-calcium diets can cause urinary calculi in wethers and rams, and inadequate calcium can lead to milk fever in lactating ewes.

Protein supplements may have plant or animal origins such as fish meal, soybean meal, and cotton or linseed meal. These supplements contain high levels of protein, but by law ruminant-derived meat and bone meal cannot be fed to other ruminants. Protein quantity is generally more important than protein quality. Livestock do not store protein except as it is used in muscle production so it must be supplied every day. Urea is sometimes used as a protein replacement, although it is not a protein supplement. Urea poisoning can occur if it is more than one-third of the total nitrogen in the diet. It should only be used in conjunction with high-energy feeds, such as corn.

Pelleted or textured feeds are produced by many feed companies as complete feeds that have been balanced for sheep of a particular age and production stage. These should not be mixed with other grain because they may cause nutritional imbalances, such as calcium-to-phosphorus ratios. They do have several advantages including complete palatability and ease of handling (fifty- to one-hundred-pound packaging), and they offer directions and guidelines for inexperienced FFA and 4-H members or sheep growers.

Numerous byproducts from the processing of traditional feed ingredients are available and are good sources of specific nutrients. These include beet pulp, corn gluten meal, soybean hulls, wheat middlings, brewer grains, and citrus pulp, to name a few. They may be economical sources of nutrients for sheep, but be aware of their total nutrient content before including them in your ration. Having them analyzed prior to use will help create a more balanced diet.

Vitamins and minerals generally need to be included in sheep rations because they need both macro and trace minerals in their diets. The most important minerals are calcium, phosphorus, salt, and selenium. Trace mineralized salt fortified with selenium will generally meet all their requirements. Complete mineral mixes that contain all the macro and trace minerals are best used during breeding, late and early lactation, and when grazing low-quality roughages. Studies have shown that selenium introduced through the diet of a pregnant ewe is better than administering selenium injections in late gestation. It is not recommended that sheep have free choice with minerals. They cannot determine which mineral is most needed and will consume the minerals that taste the best or are more palatable rather than those that may be needed.

Feed additives are products added to the ration to improve health and performance, but they do not supply nutrients. These additives include probiotics, which are living organisms of beneficial bacteria; ammonium chloride, used to prevent urinary calculi in lambs; and

Feed intake is an important consideration when formulating a ration. Intake can vary with weather conditions, health of the animals, and feed palatability. Providing shade in summer and shelter in winter are ways to help mitigate weather effects.

antibiotics, typically added to prevent enterotoxemia and respiratory diseases. These are not the only feed additives available; you should be alert to their inclusion in premixed feeds.

One final thought about feeding grain to your flock: Too much grain consumed at one time can result in a significant change in the rumen flora, which produces a large amount of lactic acid. When this occurs, the pH of the rumen drops, and this can become a fatal condition for the animal. Introducing grain into a diet at moderate increments will prevent a sudden change in pH, especially for sheep coming off pasture that didn't previously have access to grain.

Salt and vitamins are important components of a sheep's diet. They can be included in the ration or provided free choice. They can be dispensed from small containers away from the main feeding area.

Because grain is very palatable and has a satisfying taste, sheep can consume more grain than desirable if their intake is not regulated. If you slowly introduce grain into their diets, you can decrease the amount of forage to a degree. Close observation of your flock during and after feeding will keep you alert to any listless behavior that may indicate the onset of digestive problems.

FEED RATIONS

Feed rations for sheep do not need to be complicated. Many producers save money by mixing their own rations using whole grains, protein, and mineral supplements. Feed rations can be formulated using computer programs or manually calculated to provide all required nutrients for any production stage. Help with feeding rations is generally available from your county agricultural extension office. You can also hire an independent animal nutritionist to help with your program. In calculating your own rations, you may see feed cost savings by adapting to the grain market conditions. A local feed mill may be able to mix small batches so that your feed is always fresh.

One alternative, if you have a small number of sheep, is buying a prepackaged grain feed formulated especially for sheep. These tend to be more expensive but have the advantage of being purchased in small quantities that may store better in hot or humid weather.

Many feed companies can provide information or recommendations on their premixes, which are products they formulate to specific needs and package for sale. These premixes have the advantage of convenience and uniformity, but they may be offered in one brand only and they are generally more expensive.

While these products may be useful in your program, avoid products marketed as all-purpose animal feeds or generic feeds not specific to sheep. These products tend to contain too much copper, which is an essential nutrient but is one sheep are particularly sensitive to if fed in excess. Excess copper is stored in the sheep's liver and its toxicity can result in death. Use only feeds specially formulated for sheep.

ALL SHEEP RATIONS SHOULD INCLUDE:

- Water
- Vitamins
- Energy
- Minerals
- Protein

FEEDING THE RAM

Your ram is an important part of your sheep program, and keeping him in good body condition prior to the breeding season will help maintain his health and vigor. Rams can be maintained on pasture or wintered on good-quality hay before the breeding season. If the hay and grasses are of good quality, a 250-pound ram will require 6 to 8 pounds to meet his daily energy requirements. An additional 2 to 2.5 pounds of grain can be provided, along with a free choice of water and salt available at all times.

Rams can lose up to 12 percent of their body weight during a 45-day breeding season, because a vigorous ram will spend very little time eating. This equates to

Harvested wheat and oats leave behind a residue called straw. While it is not nutritious it does make an excellent bedding material. Straw is easier to move and handle if it is baled.

30 pounds of weight loss for a 250-pound ram. In many cases, forage alone will not supply adequate nutrition to keep rams in proper body condition. A body-condition scoring system can identify the best parameters for your ram to ensure optimum performance. Check with your county agricultural extension office.

WATER

Water does not have any nutritional value but is an absolutely essential part of your flock's diet. Sheep should have free-choice access to fresh, clean water. Rams and dry ewes need 2 gallons of water each day. Ewes with lambs need 2 to 3 gallons per day. Growing lambs need 0.5 gallons per day, and feeder lambs require about 2 gallons per day. Frozen water supplies, dirty drinking-water dishes, and long distances to water reduce water intake and have a negative impact on production. You can use a heated water bowl during winter to encourage adequate consumption by lambs and lactating ewes. You should check the water drinking bowls every day and clean them when necessary.

Water is often more important than feed. Your flock needs daily access to fresh, clean water at all times whether indoors or on pasture. Small tanks can be used and some sheep growers put small fish in them to reduce algae growth.

GRAZING
AND
PASTURE
MANAGEMENT

A grazing program provides a low-cost source of feed for your sheep, with a high nutritional value provided by the grasses. Forages can supply approximately 80 percent of a sheep's yearly nutritional requirements. Sheep are especially efficient in converting forages into protein and compete less with humans for edible grain crops than other livestock species. Because sheep can graze on pastures and grasses, they take less energy to produce than animals that require harvested and stored grains and forages. It is estimated that sheep are 26 percent more efficient than cattle at converting pastures and forages into marketable products. Sheep become more economically attractive as grain production costs increase.

Sheep with access to outdoor areas tend to stay healthier than confinement animals. Grazing also provides long-term benefits to the land, as well as savings in hay and manure handling and cleaner wool when compared to feedlot sheep. Grass-fed lambs can be used in a specialty marketing program. There is also the simple pleasure of seeing ewes and lambs in a grass field.

A successful grazing program also involves pasture management. As you work with your system, you will become knowledgeable about grasses, hay, and stocking rates, which are the number of animals that can be supported by the amount of available acreage. These rates are fairly easy to calculate and are important to follow so that you do not overgraze or undergraze your pastures and leave too little or too much feed for your sheep.

Stocking rates may not be important to you if you plan to purchase a substantial amount of supplemental feed, such as hay and grain, to provide the necessary forage for your flock's diet throughout the year. However, you can reduce the cost of buying extra feed by using your pastures to supply some of that nutrition.

Properly fertilized and managed pastures can provide the major portion of your flock's feed requirements through much of the year. A grazing program may range from April to early November in colder climates.

Vary the stocking rate to coincide with pasture productivity. This should result in greater plant vigor, more forage production, and less weed problems. Heavy stocking rates eventually decrease the pasture stand and forage yield, while too low a stocking rate reduces the carrying capacity and results in forage waste.

Your location will determine whether you use year-round or seasonal grazing, or whether you harvest hay during the year and store it for winter feeding. Your stocking calculations will take the size of the sheep to be pastured into consideration because mature sheep eat more hay and grain than younger, smaller feeder lambs. Although there seems to be little difference between breeds regarding their nutritional requirements, some breeds appear to have a better rate of gain from similar quantities of forages. Weather may influence these calculations as sheep growing in colder climates generally require greater amounts of feed than those being raised in warmer climates.

Understanding the quantities of feed it will take to adequately supply each animal during the year will help you plan for situations such as a dry year when there is an insufficient or no crop to harvest to meet your winter needs, or a wet year when you may not be able to harvest the crop in a timely fashion. Even with a reasonable calculation and projections, it is advisable to have a supplemental or alternative plan if you fall short on your feed supply before the new pasture crop arrives in the spring.

PASTURE PRODUCTION

A well-grown pasture of moderate density or thickness will typically produce between 2,000 to 2,500 pounds of total dry matter per acre (DM/a) in the first eight inches of growth. This kind of growth, under normal weather conditions, will produce about five tons of forage per acre during the growing season.

The recovery rate of your pastures refers to the amount of time it takes for plant regrowth or rejuvenation after sheep are removed from a particular area. The closer your sheep graze the plants to the ground, the longer it will take for the plants to recover because of the decrease in plant leaf surface area for photosynthesis. If you have sufficient pasture for a full rotational grazing program, it is best to move them from one pasture to another when they have grazed the grass down to a height of two to three inches. This will allow the plants to recover more quickly.

PADDOCKS AND STOCKING RATES

Paddocks are small enclosed areas located within the whole pasture and are generally divided with electric wire

One acre of pasture can support five ewes and their lambs during the grazing season. This calculation will vary depending on soil fertility, rainfall, pasture species, and pasture management.

to control access to specific parts of the pasture. Dividing the pasture into small areas helps manage sheep more intensively because they are allowed to eat in a specified area for a short time and trim down the growth before they are moved to another paddock to graze. Paddocks can be divided by walking the perimeter of the field to determine its total area and calculating equally sized areas. Consider developing and using paddocks to help manage your grazing system, unless you decide to allow your sheep complete access to all pasture areas.

The size and number of paddocks will be determined by the number of acres you commit to your grazing program, the health and density of your pastures, and how fast the pastures rejuvenate from each grazing session. Move your sheep through each of the paddocks quickly during times of fast pasture growth to help keep the plants from maturing too soon, which will lower their nutritional value.

The best way to determine when to move your sheep is to walk into the pasture each day and look at what has happened to the grasses in relation to the amount of time the sheep have been in that particular paddock. You will learn to read the land and grass to gain optimum benefits from your pastures.

Some areas may be left for mechanical harvest for a winter feed supply. While you have no control over weather conditions affecting pasture growth, you can determine how quickly and how much your sheep consume by controlling their access.

Stocking rates can be easily calculated by understanding the grazing capacity of your pastures. Grazing capacity is expressed in Animal Unit Months (AUM), which is the amount of forage required by one animal for one month. For this discussion, one animal unit will be defined as a 150-pound ewe that requires 2.5 percent of her body weight, or 4 pounds, of forage per day, throughout the year. This may vary between ewes but won't greatly affect these calculations.

CALCULATING PADDOCK SIZE

A simple formula can be used to determine the approximate paddock size needed to account for the daily dry-matter requirements of your flock.

(number of sheep) x (dry matter intake per day)
x (days in paddock)
÷ pounds forage/acre
= paddock size

As an example, if you have
25 ewes x 4 pounds DM x 2 days ÷ 1,000 pound/acre
= 0.20-acre paddocks (one-fifth acre)

The figure of two days used for the amount of time the sheep are allowed in a specific paddock is just an example. It does not take into consideration the very early season when the grass is starting to grow or late spring when the pastures are rapidly growing. Good-quality pastures can provide grazing for five ewes with lambs per acre. If you plan an aggressive rotation scheme, where a large number of improved pasture acres are being utilized, you can approach fifty sheep per acre. This would require substantial rest periods for the pastures after each grazing cycle. The possibilities for your farm may lie somewhere in between.

There are many options for the number of sheep you have or want to raise. The acres you commit to pastures will determine the number of sheep your land can support. Too many sheep for the available acres can lead to overgrazing and an insufficient winter feed supply. If too few sheep are grazing, they may not keep the pastures sufficiently grazed down and you will need to clip them to rejuvenate new growth.

PASTURE MANAGEMENT

The way in which you manage your pastures will be largely determined by the amount of pasture space available, number of sheep involved, and your experience. Having little or no experience with grazing does not mean you can't learn. Grazing is a time-honored, practical, and efficient method of raising sheep. Observation of your pastures and a commonsense approach can be useful while gaining daily experience with your sheep. Your pasture-management skills will increase the more you work with them. Even life-long grazers find new things to learn.

Subdivide large pastures into smaller paddocks for rotational grazing at a high stocking rate. An electric fence can be erected at a reasonable cost and is easily moved to create new grazing areas within the same field.

You may want to adjust your lambing season to coincide with maximum pasture growth periods in the spring or fall. Cool-season perennial grasses reach their maximum growth in May and June and have a second but smaller peak in the fall.

Ewes lambing in March or April make very good use of spring pastures. Move the flock quickly from paddock to paddock at the beginning of the grazing season to keep the fast growing plants from maturing too quickly.

The management of your pastures will be tied to your production program whether you raise feeder lambs, breeding ewes, or any combination. Sheep can be grazed with other animals such as cattle, horses, or goats, but each species will have different demands on the pasture used. A few cattle grazing with sheep will consume coarse stems and seedheads that sheep refuse to eat. Sheep will eat closer to manure pats than cattle and take advantage of the lush growth from the potassium, nitro-gen, and phosphorus in the urine and manure. However, there are several considerations in grazing different species in the same pasture. For example, cattle require trace minerals that typically are provided by trace mineral blocks placed with free access. These tend to be high in copper and can prove toxic for sheep if they consume too much. Some diseases can be transmitted between species, such as leptospirosis, which horses can pick up while drinking from cattle water tanks or ponds. Studying

Areas of pastures used for outdoor feeding may develop deficient growth spots. These can be inter-seeded with different varieties of grasses to promote regrowth and establish a diverse plant population.

the nutritional requirements of all the species you plan to pasture together may help you avoid problems.

The amount of time your sheep spend in any one paddock will depend on the number of sheep grazing, pasture growth, climate condition, lambing schedule, whether you graze other species with your sheep, and other factors unique to your situation. The number of sheep grazed and pasture growth will be two factors that most influence how quickly you move them from one paddock to another. In the spring, when plant growth starts slowly but accelerates quickly, you may want to move them through your system faster to restrict this growth. You can adjust your paddock size to take advantage of rapid plant growth or allow some

areas to be left to mature for mechanical harvest for winter feed.

Pasture management also includes taking an inventory of the plants growing in those areas that you want to use for grazing. Identifying the existing grasses and legumes in a pasture is important in planning improvements. There are manuals and plant-identification guides available to help you understand what is growing in your fields. This is important if you have just purchased a farm and are starting to raise sheep.

You should also have soil tests taken to determine their nutrient composition and pH level. You can take these tests yourself or have a local fertilizer company representative take them. Conducting the test yourself

Bromegrass is a good legume to include in pastures because it grows quickly in the spring and is nutritious in its early stages. However, it also quickly matures so a fast flock rotation through your pasture system will help keep the grass at a manageable level.

and sending your sample to an independent soil-testing company will ensure an unbiased result. The results can be used to determine what soil corrections need to be made to achieve optimum productivity of your pastures and hay fields. Soil corrections can turn poor-yielding fields into productive ones within a short time. Your county agricultural extension agent can provide information about soil testing as well as guides about how to read and interpret the results. Be aware that if you are pursuing an organic market there are many commercial fertilizers that do not qualify for organic production. Consult an organic representative before applying any fertilizer.

Interseeding can be one way to introduce new plant species into your pastures. This practice involves seeding in several different ways, such as drilling, where you mechanically inject the seeds into the soil with a seeder pulled by a tractor. A passive way to interseed fields is to add seeds to the grain mix and allow them to be eaten by your sheep. These seeds will travel through their digestive system, pass out in the manure, and be deposited on the ground as they move about the pastures and trample them into the soil. If you decide to use this approach, be absolutely certain that the seeds have not been chemically treated. This is a slower process than using a mechanical

seeder but may be used in pastures that have little or no access for mechanical equipment.

Another way to enhance certain plant growth is to time your grazing and mowing. Grazing heavily in the early part of the growing season when grasses come up before the legumes will favor the legumes. Grazing heavily when grasses such as bromegrass or timothy have their growing points close to the ground will retard their growth and give grasses an advantage.

Newly seeded pastures need time to develop, and you should not allow grazing sheep onto the field too early. Newly grown small grasses and legumes can be pulled from the soil by a grazing motion where the sheep tear the plants off with a movement of the head or neck. However, it is also important not to let the plants get too big or the taller grasses will shade smaller ones and slow their growth. You may have to wait to place animals in a newly seeded pasture because of the deep footprints that may develop from their movements. Walk in a newly sown pasture to determine if it will support the weight of your sheep without leaving deep hoofprints in the soil.

GRASSES FOR PASTURES

Grasses and legumes are different plant species that contain many nutrients used by ruminants and can be grown in conjunction with each other. Grasses are distinguished by plants with hollow stems, sheath-forming leaves, and minute flowers arranged in spikelets. Legumes are characterized by having more than one leaflet or leaf on their stem. Grasses and legumes are typically divided into two uses: cool season and warm season. Cool-season grasses thrive and grow in cooler temperatures and are more productive in the spring and fall. They include timothy, clover, and bluegrass. Warm-season varieties, such as alfalfa, start growing in spring but have a uniform growing pattern during the summer season. These are grasses that thrive in the midsummer heat, which normally slows down many other types. Some grasses have very good quality in the spring and can be used for pasturing if their seeds are not needed later, but they are not very palatable as they mature. These grasses include oats, wheat, and winter rye, and are sown in late fall. The growth is enough that it can be grazed before

If your pastures mature too quickly for the number of sheep you have, you can cut the pasture, allow the plants to dry, and bale and store it as part of your winter feed supply.

Lush alfalfa mixed with grasses makes an excellent pasture for grazing. Alfalfa is high in protein, has a fast recovery rate after being grazed, and can supplement a diet with little added concentrates or grain mixes.

winter. These grasses provide excellent feed in their early stages, but their nutritional quality drops dramatically after the seedheads form.

Your climate may have a significant influence on which grasses grow best in your area. There may be no single forage best suited for all farms, but some of those varieties that can be incorporated include the list shown below. The species of plants used in your pastures should be tailored to the grazing system you have designed for your region. The climate, soil type of your farm, amount of moisture your area typically receives each year, length of your growing season, and the number of sheep you want to put on your pastures are all considerations for the kind of grass and legume species seeded into your pastures.

TRADITIONAL FORAGES

Grasses	Legumes
Kentucky Bluegrass	Alfalfa
Reed Canarygrass	White Clover
Timothy	Red Clover
Smooth Bromegrass	Birdsfoot Trefoil
Orchardgrass	Alsike Clover
Ryegrass	Ladino Clover
Tall Fescue	
Bermuda Grass	
Little Bluestem	
Big Bluestem	
Buffalo Grass	

It may be better to have a mix of different grasses in each pasture rather than a single stand of one variety. Hedging your bets with a combination may prove to be worthwhile during years of abnormal weather conditions. A mix of alfalfa-brome-timothy, clover-timothy, or some other combination of grasses that grow well on your farm will typically provide sufficient forage during an average growing season.

FALL AND WINTER GRAZING

A well-managed pasture with a moderate density or thickness can produce up to 2,500 pounds of TDN per acre in the first eight inches of growth. One study on

Stockpiling refers to leaving at least five to seven inches of grass in the pasture before winter. This requires that sheep be removed from the pasture when it reaches that height until spring. Stockpiling allows you to offer your flock an early pasture feed source.

controlled grazing management has shown that gains drop when grazing removes plant material below 1,100 pounds per acre. One way to see if your sheep are consuming enough forage is to look at the pastures: bare ground does not support sheep. Another is to see how quickly the sheep lie down in the morning. If they are eating enough to satisfy their needs, the flock should be down by noon. If more time is spent eating, they are not getting enough feed.

Supplemental or winter feed will need to be used when your pastures no longer produce enough forage to meet the sheep's nutritional needs. If pastures are insufficient in the fall, you may want to consider early weaning of your spring lamb crop to reduce the nutritional demands on the ewes. However, this will affect the diets of the lambs. They will require more supplemental feeds, but they can be placed on the best available remaining pastures.

If nutrition and body condition are insufficiently increased before winter, ewes may be unable to recover body condition during the winter because of maintenance needs. A lower body-condition score will typically affect the next lamb crop.

BODY CONDITION

While body condition may fit under the nutrition discussion, it can be greatly influenced by your pastures. Body condition refers to the amount of muscle, fat, and substance possessed by the individual sheep. A scoring chart will provide guidance in determining the condition of each animal on your farm. You will gain expertise with determining the scores of your sheep the more you handle them. Each point of body-condition score is equivalent to ten pounds of weight gain.

For example, if your ewe needs to regain two points of body condition, she will need to gain twenty pounds. This will be difficult if late summer and fall pastures are insufficient to meet her nutritional needs unless supplemental feeds are provided.

Ewes will begin to regain weight about sixty days after lambing. Utilizing your pastures in rotation effectively will allow you to compensate for your ewes' nutritional needs. Adjusting your lambing schedule may have a dramatic effect on your ewes' ability to maintain or increase body condition before winter. Scheduling your lambing to coincide with maximum pasture growth periods in spring or fall will provide more nutrients to meet the ewes' needs. Ewes lambing in March or April make better use of spring pasture growth when grasses typically grow faster than those that lamb in January or February, when the ewes must be fed harvested feeds during the period of their greatest nutritional needs.

If you follow a spring lambing schedule, your fall grazing program will greatly influence the next year's lamb crop. The body condition of your ewes will influence

Pastures producing more than 1,100 pounds of plant material per acre will support weight gain of sheep. Gains drop when production is below that level. Close observation and grazing experience will help you develop an ability to determine when your pastures approach that level. *University of Wisconsin Sheep Research Station*

July and August are a good time to assess your pasture resources and decide whether or not you should consider early weaning. Heavy pastures supplied with adequate moisture may make early weaning easier while sparse pastures due to lack of rainfall may not.

Fall is a time for rebuilding the ewe flock by restoring what the ewes have put into their lambs. Your ewe flock can be used as a forage management tool and regain lost body condition for the next lambing season. *Don and Janice Kirts*

Clean pasture management and sanitation can aid in parasite control. A pasture is considered clean if it has not been grazed by the host animal for twelve months. It may also be a hay field pasture, new pasture, or pasture grazed by livestock, such as cattle or horses, that generally do not share the same parasites with sheep. Some parasites will die during winter due to freezing and thawing. *Cynthia Allen*

their ovulation rate, ability to survive cold, and future milk production. By using your pastures effectively, you can increase their body condition before winter, when it is more difficult to increase a ewe's weight. It can be done with increases in grain supplements, but this method is more expensive than achieving weight gain through grasses and pastures.

STOCKPILING

The purpose of stockpiling grass is to provide feed for sheep during the winter and in early spring before the plants begin to grow. Leaving some of the grass at a good height prior to winter, typically at least five to seven inches, will allow you to offer your sheep an early pasture feed source. Stockpiling requires sheep to be off these pastures prior to grazing them below that level. Some shrinkage will occur during the winter, but there usually will be enough for them to graze early the next year.

INTERNAL PARASITES

Sheep tend to be susceptible to internal parasites because they graze close to the ground. Many parasite larvae do not climb higher than five inches from the ground and the sheep can ingest the parasites while grazing.

Taking fecal samples to a veterinarian to test for parasites will give you an idea if you need your ewes dewormed. Samples taken before and after treatment will usually indicate whether or not a treatment was successful.

The best way to manage internal parasites is to practice clean pasture management and sanitation. A clean pasture is considered to be one that has not been grazed for twelve months. It may also be a cut hay pasture, newly seeded pasture, or pasture grazed by other livestock such as cattle or horses that generally do not share the same parasites with sheep. Some parasites die off on pastures during the winter due to freezing and thawing; however, snow cover insulates the larvae. An effective program for dealing with internal parasites will be discussed in Chapter 10. But these mentioned practices may be employed as a way to handle mild infestations.

If you are following an organic protocol, you should be aware that parasiticides are generally not allowed for regular use in organic production. You will need to use pasture management to control internal parasites.

BENEFITS OF GRAZING

Grazing has many environmental benefits besides providing forage for your sheep program. Pastures and hay fields are the best protection against soil erosion and runoff. Organic carbon becomes trapped in these soils and may help reduce atmospheric levels of carbon dioxide and greenhouse gases. The level of perennial grasses within the plant community tends to increase. This has been shown to increase water infiltration and decrease erosion.

Grazing may allow you to control grass, weeds, or small brush in areas not accessible to machinery, such as on rocky land, hillsides not conducive to safely handling machinery, and areas where machinery cannot be turned around. In many cases, you may be able to pasture around buildings for weed and grass control, keep fence lines clear, or keep lanes and trafficways clean. Sheep will be able to reach areas you may not.

Pastures become habitats for many grassland bird species and other wildlife. Grazing can also influence the plant diversity because of your interseeding program that may have beneficial effects on useful insect populations. As the manure decomposes on the field, it provides nutrients for the plants and homes for insects that feed on the fibers and organic materials. Insects provide food for grassland birds that may enhance your farming experience.

Public concerns about humane treatment of animals and food produced under natural conditions can be addressed with the use of pastures. Pastures allow sheep a chance to pursue their natural instincts, reduce their stress level with more space to freely move about, and increase their level of comfort.

Sheep graze close to the ground, which makes them more susceptible to internal parasites than other farm animals because many parasite larvae do not climb higher than five inches on the plants. Rotational grazing can help reduce internal parasite infestation of your fields. Testing fecal samples can indicate the parasite load of your sheep. Develop an effective deworming program with your veterinarian if treatment is needed.

BREEDING AND REPRODUCTION

A successful reproductive program will enable you to replenish your flock and keep the number of sheep on your farm at a sustainable level. By understanding the natural reproductive processes, you will be able to utilize the animals you have chosen, regardless of breed, to repopulate your flock even after selling some over the course of a year. The reproductive performance of your flock may not be as much of a concern if you are in a production system where you purchase lambs and only raise them to market weight, but it is still fundamental to any financial success. By achieving high reproductive performance, you will be able to maintain your breeding flock and have more lambs to raise or sell.

REPRODUCTIVE ANATOMY

Sheep have a similar reproductive structure to other mammals. The position of the reproductive organs serves the same functions. There are many resources available for you to consult to increase your knowledge of ram and ewe reproductive systems.

ESTRUS

The onset of estrus, commonly referred to as heat, establishes sheep puberty. This first signal of sexual maturity is influenced by age, breed, weight or body size, nutrition, and season of breeding. Most ewe lambs will reach puberty between five to twelve months of age and will typically reach it by their first fall. This is the reason why spring-born lambs tend to exhibit puberty before fall-born lambs. Because lambs born early in the season tend to reach puberty earlier than those born late in the season, it is one reason to maintain a compressed breeding season.

Other factors influencing puberty include the amount of feed given to pre- and postweaned lambs. Single lambs have a size advantage over twins or triplets. Meat and hair sheep breeds reach puberty earlier than fine- or coarse-wool breeds. Crossbred lambs tend to cycle earlier than purebred lambs, which is most likely the result of their hybrid vigor.

The female reproductive system has two ovaries that lie within the pelvic cavity of the ewe. The ovaries are similar in size and shape and are attached to the tissues that envelop most of the other reproductive organs. When the ewe is not pregnant, the ovaries lie within or on the front edge of the pelvic cavity. In advanced pregnancy, the ovaries are carried forward and downward into the abdominal cavity along with the enlarged uterus containing the lamb or lambs.

The ovaries set the pace for the rest of the reproductive tract that must be adjusted to receive the fertilized egg and carry the developing fetus through to birth. Ovaries have two functions: the production of eggs or ripe ova and the secretion of hormones. The hormones cause necessary adjustments in other parts of the tract to take place. The heat cycle, or estrus, is divided into several well-marked phases.

Because most sheep are seasonal breeders, the first phase is set in motion by the declining daylight pattern in fall. This development period, or proestrus, will govern the initiation and release of hormones to begin the cycle. As the light period decreases, a signal is sent to the pituitary gland in the base of the brain that secretes follicle-stimulating hormone (FSH) and luteinizing hormone (LH) to initiate and stimulate the ovary's activity to start the process. FSH causes small follicles, called oocytes, to grow within the ovary, principally due to an increase in fluid. The oocytes appear on the surface of the ovary and resemble a water blister.

Each follicle contains an egg, and considerable amounts of hormones, most notably estrogen, are produced as it develops. The follicles continue to grow and the eggs begin to mature as the estrogen levels increase. The elevated estrogen levels in the follicle lead to increased

Replenishing your flock should be one goal of your program. This will be influenced by ram selection, nutrition, and the health of your ewes. Take time to learn about the factors that will have an effect upon the reproductive performance of your flock.
Cynthia Allen

estrogen levels in the blood. When the estrogen concentrations in the blood become high enough, the ewe shows signs of heat. The estrus or heat stage follows, which is the period of desire and acceptance of the ram.

The follicle on the ovary ruptures during the latter period of the heat stage. Estrus will usually last twenty-four to thirty-five hours but can range from twenty to forty hours. Ovulation typically occurs twenty to thirty hours after the initiation of estrus.

After the follicle has ruptured, the egg travels down the oviduct where it is fertilized with sperm in the isthmus of the oviduct. The former follicle cavity forms a corpus

Sheep are seasonal breeders and ewes are stimulated to cycle by the declining daylight pattern in autumn. You may be able to develop a lambing program that occurs three times a year and extend your marketing options depending on the breed you choose to raise. *Nikole Riesland-Haumont*

luteum (CL) that will produce and secrete progesterone to prevent further secretion of FSH and LH by desensitizing the anterior pituitary to gonadotropin-releasing hormone (GnRH), which stimulates the release of FSH. The CL will persist on the ovary throughout pregnancy and will suppress further heat periods from occurring and interfering with the follicle process already in progress. The high level of progesterone causes the ewe to be uninterested in the ram.

After fertilization occurs, the egg implants into the lining of the uterus and the fetus develops. If fertilization fails to occur, the corpus luteum regresses in size and loses its influence on the process. This allows the GnRH levels to increase and stimulate the release of FSH as the whole cycle starts over again. The ewe will again exhibit signs of estrus approximately seventeen days later. The estrus cycle for sheep has an average of seventeen days, but it may range from fourteen to twenty days.

Most pure- and crossbred sheep will normally ovulate one, two, or three eggs. If two eggs are released during ovulation and both are fertilized, twins will result; if three eggs are fertilized there will be triplets. The number of eggs released is also controlled by nutrition and the environment. The ovulation rate for all breeds is at its maximum around the shortest day in autumn and may remain there for up to sixty days. Because of these natural conditions, most sheep breeding in the United States occurs between August and November when fertility is highest and most efficient. Ewes bred at this time generally produce the highest percentage of multiple births. This cycle can be manipulated, synchronized, and altered by the use of hormones to shift the estrus periods in different groups of your flock or to consolidate ewes into different lambing patterns. Your local veterinarian should be able to provide information on different options.

HEAT DETECTION

The signs of heat or estrus in sheep are not as pronounced as they are in cattle or pigs. Estrus usually cannot be detected unless a ram is present. When mature ewes are in heat, they may seek out the ram and stand for mounting. Other signals include wagging their tails vigorously, nuzzling the ram, or trying to mount him.

The introduction of a ram near the end of the prebreeding period may help mentally stimulate the ewes and bring about earlier ovulation and estrus activity. Rams can be kept in an adjoining pasture to the breeding flock about ten to fourteen days before breeding begins. Rested fertile rams can be introduced at the beginning of the breeding season.

Ewe lambs and yearlings are normally shy breeders and should be separated from older ewes at the time of breeding. In some cases, it may be better to use rams of smaller breeds on young ewes to minimize the chance of lambing difficulties. Ewe lambs should not be bred until they reach about 70 percent of their mature weight. For example, if the mature weight of your breed is 160 pounds, then ewe lambs should not be bred until they weigh at least 112 pounds.

Breeding ewe lambs increases their lifetime productivity but may not be economical for all producers.

Heat detection in sheep is different from cattle or pigs where a mounting posture is exhibited with other animals. Ewes do not mount other ewes. Having a ram in close proximity to the ewes will increase their mating response.

In some instances, ewe lambs may have more lambing problems the first time. The flock may improve in its capacity for ewe lamb breeding, which can be a sales factor to stress when selling breeding stock, if replacement ewes are chosen for their ability to breed as lambs.

PREGNANCY LOSSES

There are several factors that can lead to pregnancy mortality. High temperatures are detrimental to fertility, embryo survival, and fetal development. High temperatures during the first nine days after fertilization can also reduce conception rates. Heat stress during gestation can impair fetal development and cause lambs to be significantly smaller at birth. High temperatures may also lengthen the estrus cycle by one or two days and can depress estrus behavior toward the ram, so it is important to provide adequate shade during this period.

Eggs ovulated as singles generally have higher survival rates than eggs shed as pairs. Twinning may be beneficial if you are trying to improve reproductive rates, provided management and nutrition are adequate.

In unusual instances, pregnancies may be lost because the ewe doesn't recognize she is pregnant. The hormonal signals may become confused and cause the

A marking harness can be strapped to the chest of a ram during the breeding season to identify which ewes have been bred. As the ram mounts the ewe, the color crayon attached to the bottom of the harness leaves a mark identifying the ewes that have been serviced.

CL to regress, resulting in the depression of progesterone, which results in the embryo's death because progesterone is important to the maintenance of pregnancy. These ewes may be difficult to identify at first but persistent heat cycles should alert you to the fact that something abnormal is happening to this ewe.

Most embryo losses occur during the first thirty days after fertilization. Embryos generally implant between days twenty-one to thirty, so anything that may create a disturbance, such as shearing, vaccinating, or pronounced shifts in handling or feeding practices, should be avoided during this time.

Pregnancies can be determined by a ewe's failure to return to heat or technologically by using ultrasound scanning procedure. An ultrasound scan can be done successfully thirty-five to sixty days after breeding. If you choose this management tool, it should be done by a qualified technician.

GESTATION PERIOD

The normal gestation period of a ewe is approximately 147 days, but it may range from 144 to 152 days. Medium-wool breeds and meat-type breeds typically have a shorter gestation period than the fine-wool breeds. Early-maturing breeds tend to have shorter pregnancies than the late-maturing breeds. Ewes carrying multiple fetuses tend to lamb earlier. The gestation period may be shortened two or three days because of external factors such as high temperatures or nutrition levels. During the first fifteen weeks of pregnancy, ewes need nutrition levels that account for body maintenance and some fetal growth. The last four to six weeks of gestation, when the majority of fetal growth occurs, is the most critical period for ewe reproduction. This places a greater nutritional demand on the ewe. You must provide an adequate diet to minimize the chances of pregnancy toxemia and milk fever, which can result from inadequate diets.

Ewe lambs will require extra nutrition while pregnant because they are still growing. They should be separated from older ewes because they need a different diet. In addition, ewe lambs are unable to compete well at the feed trough and won't gain the proper amount of weight.

PREBREEDING AND FLUSHING

If you have experience with your breeding flock, you can practice some selection of your ewes to ensure a more successful breeding season. Selection includes culling ewes with a history of not raising their lambs, poor body condition due to old age or missing teeth, a history of prolapsed uteruses, udders with lumps or abscesses, feet and leg problems, or anything else that may keep ewes from becoming a productive flock member during the coming year.

Flushing is the practice of increasing the nutrition level of your ewes so they are in a weight-gain situation to prepare for breeding. Flushing can be accomplished by supplementing the usual summer diet with grain or

a better pasture. It works best when this practice begins at least twenty-one days prior to turning the ram in with the ewes to be bred because this will allow them to pass through one estrus cycle. Ewes typically will shed more eggs in the second heat and are more likely to have twins if bred during this second cycle.

The goal of this practice is to get the ewes in better physical condition for breeding, but it also helps synchronize them into heat so the breeding and subsequent lambing sessions are not spread out over a long period. Flushing may also have an effect on twinning. The U.S. Department of Agriculture (USDA) estimates that flushing can result in an 18 to 25 percent increase in the number of lambs.

Begin your flushing program by providing a quarter pound of whole or cracked corn a day per ewe. Moving them to a fresh pasture may also enhance the potential for ewes to respond to flushing. Increase the grain amount to reach three-quarters of a pound each in the first week and continue that level for the next seventeen to twenty-one days. There doesn't appear to be an advantage to

The rump of the ewe is marked with the color crayon on the ram's harness during mating to make it easy to identify which ewes have been bred and which still need to be bred. Daily observation of the ewes during the breeding season is essential when using a marking harness to record those serviced.

starting the flushing earlier than this. This practice can be continued for about thirty days after breeding with decreasing amounts of feed. Be careful not to put breeding ewes on clover pasture because it contains estrogen and lowers lambing percentages.

Breeding ewe lambs do not need to be flushed because they will not have reached full size by lambing time. They should be bred near the end of the breeding season, such as September or October, instead of August. The result of all your work and that of the ram is a flock of pregnant ewes that need sufficient nutrition for the next five months to successfully birth healthy lambs.

THE RAM

Your ram is an integral part of your sheep-raising program. A ram must be vigorous and fertile. Infertile, diseased, or disinterested rams often cause poor lambing rates and a significant decrease in flock profitability.

The "ram effect" is a viable consideration when planning your breeding program. The presence of a ram, especially the smell, has a great effect on the onset of estrus activity in ewes. However, this stimulus is not as pronounced if the ram has constant contact with the ewes. The ram should be placed in an adjoining pasture about two weeks ahead of the time breeding season starts.

Rams should be isolated from ewes for at least six weeks for the ram effect to work. Once the ram is placed with the ewes, he can stay for fifty-one to sixty days and be able to mate with all the ewes, even those that repeat an estrus cycle and yearlings that may be late in coming into heat. The greatest benefit of the ram effect is the synchronization of the heat cycles that will result in large numbers of ewes ovulating, conceiving, and lambing in a relatively short period of time. You should have several young, healthy, well-grown rams available at breeding time in case problems arise with the one being used.

A healthy, well-grown ram lamb weighing seventy-five to ninety pounds can service up to thirty ewes. A fit yearling or mature ram can cover forty to fifty ewes or more and can usually mate with three to four ewes per day without any noticeable effects during the breeding season. These rates may depend upon season, temperature, sex drive, and body condition. Other factors that may affect ram performance are malnutrition,

internal parasites, or disease that depresses his desire to mate or causes infertility. Common diseases such as foot rot or a disease that affects any of the external breeding organs can make it impossible for a ram to successfully breed ewes. Follow a regular parasite-control program and vaccination schedule to minimize these negative effects.

Rams can vary in their sexual behavior and may repeatedly mate with the same ewes while others are in heat at the same time. A ram may lose up to 15 percent of his body weight during the breeding season. It is important to maintain a good nutritional program for your ram during this time. Daily observations of him will alert you to any problems.

If you are only using one ram, you may want to consider a fertility test to determine his semen quality and mobility. Semen testing by qualified veterinarians can identify rams that may not be able to adequately impregnate your ewes. If semen testing is not possible, the use of a marking harness can be beneficial. A marking harness is a holder on his chest for a marking crayon or paint. Each ewe is marked with the color of the day he breeds her. Using one color for the first sixteen days and another color for the next cycle will allow you to determine which ewes returned to heat and those that didn't and will most likely become pregnant. This will allow you to determine which ewes should lamb on a particular date. If many ewes are being remarked with a second color, you may have a low-fertility ram.

You may be able to raise a ram from your own flock that will have an acceptable growth rate and be some indication as to his offspring's rate of gain. The way a ram is raised may have a significant effect on his sexual performance. Rams raised in an all-male group tend to show lower levels of sexual performance later in life and some will exhibit no sexual interest in receptive females. The ram lamb you select for later use should be raised with ewe lambs or in a mixed group until he reaches puberty.

If rams are raised in a mixed group, social dominance will be demonstrated between males. This may be a good thing if the dominant male is the ram you want to use for breeding purposes. The mating success of a dominant male far exceeds that of a subordinate one.

However, this may cause problems since aggressive rams can pose a hazard to the family if not handled properly.

One caution when raising a ram for breeding: Do not make him your family pet. This can lead to a butting behavior, which is extremely difficult to curtail once it is learned. A butting ram can inflict serious injury, especially to children. Proper handling of the ram when moving about in the pastures or pens will minimize the potential for injury or confrontation.

ARTIFICIAL INSEMINATION

Artificial insemination (AI) is possible in sheep but is not commonly used. While AI is relatively easy in cattle and swine, it is more difficult to perform in ewes because of their complicated cervix. Sheep also fail to show visible signs of heat like cattle, such as mounting other cows, which makes it more difficult to identify the correct time to inseminate the ewe.

There are four methods of artificial insemination for ewes including vaginal, cervical, transcervical, and intrauterine. Vaginal AI is the simplest form and involves depositing fresh semen into the vagina without any attempt to locate the cervix. Results vary greatly and this method is unsuitable for use with frozen semen. Cervical AI is cheap and relatively easy to accomplish. The cervix is located, usually by using a speculum fitted with a light source, and the semen is deposited into the first fold of the cervix. Conception rates with fresh or chilled semen are relatively good but are much lower with frozen and thawed semen. The transcervical AI method involves grasping the cervix and retracting it into the vagina with a pair of forceps to allow an inseminating instrument to be introduced into the cervical canal. The intrauterine method bypasses the cervix and deposits semen directly into the uterine horns using an endoscope that allows the technician to see the reproductive tract and inject semen directly into the lumen of the uterus. Conception rates are about 50 to 80 percent.

AI has not been as effective in advancing sheep genetics as it has for other species, such as cattle. Although laparoscopic AI pregnancy rates are similar to those achieved with natural service, it takes more time and expertise to accomplish than using a good, healthy ram.

RAM EPIDIDYMITIS

Epididymitis is a sheep disease that occurs in most of the sheep-producing countries of the world. In the United States it is more prevalent in the western states than other areas. It is characterized by both acute and chronic inflammation of the testicle and epididymis of the ram. This disease results in economic losses for producers because of reduced fertility, shortened breeding life, expensive immunization, and increased labor. The most significant economic impact is a lowered conception rate, extended duration of the lambing period, failure of a high percentage of ewes to conceive, and occasional abortions and weak lambs.

Five different bacteria may be involved with this condition that affects the epididymitis, prepuce, and penis of young rams. Manual palpation of the testicles and epididymis is important to determine the presence of scar tissue at the tail of the epididymis where it attaches to the testicle. Because a ram may be infected months before lesions appear, a serum Enzyme-Linked Immunosorbent Assay test (ELISA) may be required to determine its presence and control this disease. Another method of diagnosis is a semen culture.

Epididymitis is transmitted in several ways including from an infected ram to a ewe to another ram. It may also be contagious from ram to ram or through contaminated feed and water. There is no effective treatment for rams with clinical epididymitis. It can be diagnosed from lumpy tissues or swelling of the epididymis. Rams infected with the disease should be culled. Mixing infected rams with clean rams should be avoided. The best control is to test your rams and cull any suspected of having this disease. A discussion with your local veterinarian can help you explore options should you find an incidence of this disease in your flock.

BREEDING PROGRAMS

There are several different breeding programs you can use with your flock including linebreeding or inbreeding, crossbreeding, outcross breeding, and grading up. Inbreeding mates sheep that are closely related to increase the frequency of pairing similar genes. The physical expression of these genes will produce sheep more similar in certain traits than the normal population. As an example, inbreeding involves the mating of sire to daughter, dam to son, or brother to sister. Inbreeding generally helps develop distinct family lines and is useful when crossing with other unrelated family lines. However, inbreeding tends to lower reproductive efficiency, survival rates, and growth rates.

Linebreeding is a mild form of inbreeding and can be successful if used wisely. The objective of linebreeding is to keep a high degree of relationship between sheep in the flock and some outstanding ancestor or ancestors. In contrast to inbreeding, which generally does not attempt to increase the relationship between the offspring and any particular ancestor, linebreeding deliberately attempts to keep that relationship. An example of linebreeding is when a granddam is bred back to her grandson.

Crossbreeding involves mating sheep of two different breeds to utilize the heterosis or hybrid vigor, which is the increased hardiness and growth performance gained when crossing two different breeds. Heterosis takes advantage of the best qualities of each breed and offspring generally are more efficient in growth and health. However, the opposite may result where the lamb is a combination of the poorer qualities of the parents. The result of crossbreeding is generally more positive than negative.

Outcross breeding is the use of a ram from the same breed that does not have any relationship to any other sheep in the flock. Outcross-bred animals will exhibit the same general characteristics as the rest of the breed but have no close genetic relationship to each other. An example is when two mated individuals

Ewes that do not settle from a service will return in heat in about seventeen days. You should keep your ram with the ewes you want bred for at least sixty days to help ensure a maximum number of pregnancies.

have no common ancestor within the last three or four generations.

Grading up is using a purebred or exceptional ram on a flock of ordinary ewes, keeping the best offspring, and culling the rest. If you use this method for several years and keep the best of the resulting lambs and discard the original ewes, you will steadily improve the quality of the breeding flock. Your success with this approach will depend on selecting a good ram and selecting the best replacement ewe lambs.

There are many different breeding combinations and any successful sheep producer will keep extensive records to track progress and identify trends in his or her breeding program. Each breeding system has unique features, advantages, and disadvantages. Having a basic understanding of these breeding systems may help you determine the method you wish to incorporate into your sheep-raising program.

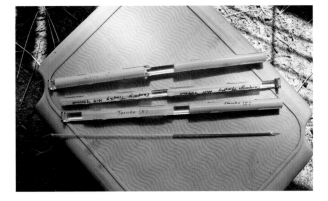

Artificial insemination has proven to be a good way of incorporating superior rams into a breeding program through frozen semen. The semen is packaged in ¼ cubic centimeter plastic straws and transported in liquid nitrogen tanks. One straw or semen unit (shown at the bottom in pink) is used for each insemination, which should be done by an experienced technician.

SHEEP MANAGEMENT

Identification of your sheep is one of the first steps to good management, especially if individual markings are similar. As your flock numbers increase, so does the need for accurate identification records. These can be used to track individual animals, families, or the entire flock.

Sheep management involves a varied set of skills and knowledge. It involves decision making on feed rations, culling, replacement animals, treatment options and protocols, recordkeeping, mating systems, and many other practices that influence your sheep-raising program.

Sheep need day-to-day care and attention even though they may be self-sufficient in many ways. Daily care includes observing their physical state and condition.

Successful sheep growers develop many and varied skills that help them handle problems when they arise. Good sheep management often takes a common sense approach to address many health concerns. *Sue Carey*

It also includes anticipating potential problems and taking steps to eliminate them as quickly as possible. Simple things, such as fixing broken or downed fences, can be taken care of before they become areas of escape by sheep or entry points for predators. Health concerns, such as illnesses, lameness, or other physical ailments, will need your prompt attention. Your ability to anticipate problems will come with experience and close observation.

You will have to make daily observations of your flock, particularly during extreme weather and lambing time. Problems with any species of livestock can surface quickly; your response to these critical events may determine whether a ewe or lamb lives or dies and directly affects the profitability of your program. Raising sheep is a business, as well as a hobby. Like most successful businesses, managers who pay attention to the minute details often have more success than those who don't.

Sheep growers have the ethical responsibility to provide their flock with conditions where they have access to quality feeds, can reside without fear, and are treated in a gentle and humane manner. It is in your best interests, financially and philosophically, to have sheep that are content and easy to work with.

INDIVIDUAL EWE RECORD

EWE IDENTIFICATION:		SCRAPIE ID:		BREED:
BIRTH DATE:		SIRE:		DAM:
TYPE OF BIRTH		BIRTH WEIGHT:		WEANING WT:

COMMENTS:

DATE LAMBED	SIRE	SEX OF LAMB	BIRTH WEIGHT	TYPE OF BIRTH BORN AS	TYPE OF BIRTH RAISED AS	LAMB ID	WEANING DATE	WEANING WEIGHT	COMMENTS

Recordkeeping is an important part of flock management. It allows you to note lambing information, parentage, illnesses, treatments, and many other pieces of information that may determine the retention or removal of a particular animal. This example of an individual ewe record contains a lot of information. *Department of Animal Sciences, University of Wisconsin-Madison*

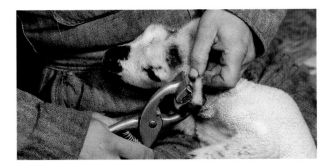

An ear tattoo is a permanent identification method that uses indelible ink embedded in the ear. A tattoo stamp contains a code that denotes birth year and individual identification. *Department of Animal Sciences, University of Wisconsin-Madison*

HEALTH MANAGEMENT

Animals can get sick even with the best of care. The key is to minimize the number and severity of those illnesses to the greatest extent possible. The first step in minimizing the effects of a sickness is to visually inspect your sheep on a daily basis. This does not necessarily require a great amount of time, depending on your flock size, but walking out to where the sheep are and looking for signs of any abnormalities in health, movements, alertness, appetites, or other conditions will help you keep abreast of their condition. You will be able to identify sheep that appear listless or limp, have droopy ears, or that may not be eating normally, which can indicate a problem. Identifying problems will allow you to move quickly to address them. Any sheep that exhibits symptoms of illness will generally not overcome it without your intervention or help. While you may be able to perform some health practices, a licensed veterinarian is required to administer some vaccinations, medical euthanasia, and certain antibiotics or steroids.

Depending on your program, philosophies, and goals, there are three treatment systems that can be used on your sheep: conventional, homeopathic, and herbal. Each has advantages and disadvantages. Understanding the differences may help you decide which method to use in your program.

TREATING ILLNESSES

You do not need a veterinary-school education to treat many of the sheep illnesses that may occur in your flock. Experience is the best teacher, but you can gain insight and understanding by reading about sheep health and the practical application of administering treatments.

UNIVERSITY OF WISCONSIN MADISON
ARLINGTON SHEEP UNIT
2008 SPRING LAMBING RECORDS

EWE LAMBS—LEFT EAR

RAM LAMBS—RIGHT EAR

Date	Dam	Sire	Lamb ID	Sex	WT	GT	Disposal	Date	Moved	Comments

Lambing Record Chart: This type of chart can be used to record information during the lambing season. It is important to record the disposal of all sheep from your farm whether they are sold to market or other buyers, or because of death. Recording the purchaser's name and address will make it easier to trace the movement of your animals should that information be needed later. *Department of Animal Sciences, University of Wisconsin-Madison*

BARN RECORD

Date lambed	Dam	Sire	Sex of lamb	Birth Type	Birth Weight	Lamb ID	Comments

Barn Record: Keeping production records will help monitor flock performance. Providing comments about lambing ease, vigor of the lambs, and mothering ability of the ewe may be useful information in making replacement decisions.

There will be times when you have no choice but to call a veterinarian to help with a particular health issue. However, the quickness of a problem surfacing or the severity and extent of the problem may have to be solved by your own initiative, such as during a case of bloat, where time for resolving the problem is extremely short and a call to a veterinarian most likely will be too late.

If you administer medications, you should pay close attention to where injections are given because needle marks left on the carcass of market animals may lower their price. Oral medications may be mixed with feed, syrups, or water. Subcutaneous injections are usually given in the loose skin where the neck and shoulder join. If intramuscular injections need to be given, they should be done in the neck muscles. You should read the labels of all products used on your sheep and follow the directions indicating the amount of time needed to withhold the animal from market or slaughter.

It is important to keep accurate treatment records for your sheep because it is difficult to remember the exact details of a single treatment six weeks later. Good records also help identify recurring problems that may need to be addressed by changes in management, treatment, or the removal of an animal with a chronic problem.

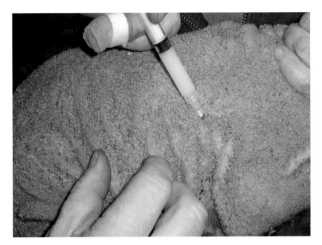

Ear tags are a simple form of individual identification and can contain useful information about the animal including birth year, sire, and flock number. Different color tags can be used for different birth years or sires. Neck chains should be avoided because they can snag sheep to obstacles and are not considered a form of permanent identification.

Viral illnesses and vaccinations are typically handled through injections with a sterile syringe. There are many shapes and sizes of syringes and needles and you should become familiar with their use. Needles should be properly disposed after each use to reduce the risk of infecting another animal. Be sure to read and understand all label directions and withholding times for the products.

ARLINGTON SHEEP GROUP TREATMENT FORM 2008 SPRING LAMBING RECORDS

DATE PERFORMED:

GROUP DESCRIPTION

RECORDED:

TREATMENT(S) PERFORMED:			PRODUCTS GIVEN:		DOSAGE	MODE
1								
2								
3								
4								
5								

Group Treatment Form: Maintaining accurate group records is essential for identifying which animals have received treatments and the date the treatments were given. They can help identify recurring problems within your flock and prevent accidental double dosing. *Department of Animal Sciences, University of Wisconsin-Madison*

ARLINGTON SHEEP
ANIMAL HEALTH/TREATMENT RECORD

ANIMAL ID:

GROUP DESCRIPTION   DATE:

SYMPTOMS/DIAGNOSIS/HISTORY:

TREATMENT—INDICATE ALL SUBSTANCES GIVEN, (DOSAGE, METHOD OF ADMINISTRATION) OR PROCEDURES PERFORMED

COMMENTS:

DATE	TIME	BODY TEMP						EMPLOYER INT

Individual Treatment Record: Animal health and treatment records should be kept for each member of your flock. It is important to record any administered medications so the withdraw times are strictly observed and animals do not enter the food chain before the medications clear their systems. *Department of Animal Sciences, University of Wisconsin-Madison*

CONVENTIONAL TREATMENT

A conventional treatment program for sheep typically involves the use of antibiotics to relieve the symptoms of the illness or disease. Although antibiotics may correct the problem in the short run, residual effects may affect marketing those treated animals. You will need to fully understand the withdrawal time for the antibiotic used and be aware that different products have different withholding times.

Antibiotics were first used to treat disease in food-producing animals during the mid-1940s. By the early 1950s, the livestock industry saw the introduction of antibiotics in commercial feed for general use in sheep. Since the 1960s, antibiotics have served three purposes: as a therapy to treat an individual illness; as a prophylaxis to prevent illness in advance; and as a performance enhancement to increase feed conversion, growth rate, and yield.

Bacterial diseases, if unchecked, generally cause pain and distress in an animal and result in an economic loss. Antibiotics can be used to help reduce this suffering and distress and speed the recovery of the infected animal. When used responsibly, antibiotics can be an essential element in the fight against sheep diseases. In rare instances they may be used to prevent some diseases that might occur in a flock if a high probability exists for most or all of the sheep becoming infected.

The use of antibiotics, whether in a therapeutic protocol and especially on a routine basis, should be

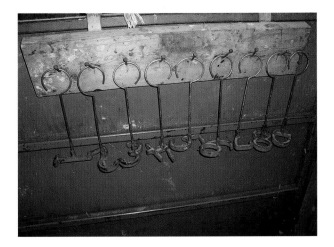

Temporary identification is sometimes useful in a flock. Paint branding, marking crayons and sticks, and sprays will generally last several weeks. Special marking sprays are available. You should never apply a household spray paint to sheep.

approached with caution to prevent overmedication, residues, and bacterial resistance. The overuse of antibiotics in animals entering the food supply continues to be a major concern of consumers and should also be a concern to producers.

Familiarizing yourself with the diseases that may require antibiotic assistance will help avoid unnecessary usage. A discussion with your local veterinarian may provide insights that will be useful in your program.

ALTERNATIVE TREATMENTS

The concern about residual antibiotic effects in animals and humans has driven interest in alternative treatment programs. Information is available for homeopathic treatment systems, and this approach is ideal for those considering organic, sustainable, or biological farming methods.

Homeopathy is based on the idea that bacteria are not necessarily bad and do not need to be destroyed. The sheep's reaction to the illness, not the illness itself, is treated. Homeopathic treatment involves the natural stimulation of the sheep's immune system so it can fight off the bacteria that might otherwise cause a disease. Homeopathy can be used to support the development of a lamb's immune system before it is even born by work-

ing with the pregnant mother. The first two weeks of a lamb's life are the most important and will determine, to a large extent, its health later in life. A sickly lamb does not become a healthy adult overnight, but a healthy, vital lamb can have the ability to stay healthy.

A homeopathic treatment is called a remedy and is administered in small dissolvable pellets. Remedies are derived from all-natural sources, including animal, mineral, or plant, and are prepared by a qualified homeopathic veterinarian or pharmacist. A remedy is given in doses usually marketed in specific amounts and based on a system of potencies. In tincture form, the remedy is added to a sugar-granule base and allowed to soak, after which it becomes stable and can remain active for months or several years if properly stored.

Remedies can be purchased in multigram vials and usually contain sufficient amounts for several administrations for one or several animals. A dose may consist of several small pellets given at one time, depending on the condition being treated, and may have several applications during one day. Application methods typically require that the dose is placed directly on the lamb or ewe's tongue to be dissolved by saliva. The granules do not need to be swallowed because

Handling sheep in an easy, humane manner is good management. People, especially children, can become unnecessarily injured because of a lack of good safety. A chute system is an easy way to handle and move sheep that is also less stressful to them. *Department of Animal Sciences, University of Wisconsin-Madison*

A chute system allows a systematic approach to working with sheep. These are useful during times of vaccinations, shearing, or other management practices. This chute system funnels sheep in a manner that keeps them moving forward but without crowding. The head gate shown here allows individual sheep to be restrained while you are working with them.

homeopathic remedies can be absorbed through the palate or tongue. The potencies are determined by the dilutions made from the refined crude product. This refinement develops the inherent properties of the remedy used for the specific needs of the sheep. Homeopathy is not a substitute for preventive measures like good nutrition, air quality, or proper sanitation.

HERBAL TREATMENTS

Herbal treatments are another alternative that uses remedies based on preparations made from a single plant or a range of plants. Applications are by different routes and methods depending on the perceived cause of the disease condition. These applications can be made from infusions, powders, pastes, and juices from fresh plant material. Topical applications can be used for skin conditions, powders can be rubbed into incisions, oral drenches can be used to treat systemic conditions, and drops can treat eyes and ears. Information is available from alternative health stores or books published on these topics to help explain the products and their usefulness.

DISEASE PREVENTION

Many health issues have the potential to affect the performance of your sheep and their lives. Good planning, sanitation, and management, along with use of common vaccines, will generally help avoid most disease problems. Find a local veterinarian who treats sheep and consult him or her about a flock health program. A list of veterinarians in your area or region is usually available from your county agricultural extension office. You may want to consult a veterinarian prior to bringing any sheep onto your farm.

There is no substitute for good sanitation. This simple practice will help safeguard your flock from many diseases. While all areas of your farm can benefit from good sanitation, it should particularly be directed toward your lambing areas as a first line of defense. Disinfectants have little or no effect in areas that contain manure and dirt because the organic material in them will inactivate many of the ingredients. Lambing pens should be cleaned and disinfected between ewes using the area. Before placing a ewe in the pen, wash her udder with detergent and water. This will help prevent the newborn lambs from swallowing infective materials with their first meal. Other practices that can be used to prevent the spread of some diseases include immediately isolating sick ewes or lambs from the rest of the flock and not allowing other sheep access to that area until it has been thoroughly cleaned and disinfected. Being dependent upon drugs to control sheep diseases is a poor substitute for balanced rations, sanitation, and sound management aimed at disease prevention.

INTERNAL AND EXTERNAL PARASITES

One important management practice regarding sheep is controlling internal parasites. Parasites can live inside a sheep or outside, under the wool. Compared to other animals, sheep are more resistant to bacterial and viral diseases but more susceptible to internal parasites.

Rams should be viewed with caution at all times no matter how tame they may seem. A large ram, especially one with horns, can be dangerous to young children and other family members. *Cynthia Allen*

Sheep that are weakened and run-down from extensive infestations of internal parasites are more susceptible to diseases, which makes control of internal parasites an important management decision.

Common internal parasites are liver flukes, tapeworms, large stomach worms, and brown stomach worms. An internal parasite program can be effectively employed by using certain anthelmintics or pharmaceutical products. These should be given on a routine schedule and you should keep accurate records of each application and which sheep received it. Ewes should generally be given an anthelmintic prior to breeding, before or at lambing, and one time during the summer. Lambs should be given an anthelmintic at weaning and at six- to eight-week intervals, depending on the level of infestation.

To determine which anthelmintic products should be used, you will need to assess what parasites are present. This can be accomplished by having a state veterinary diagnostic laboratory check the feces for parasite eggs to determine the worms present and degree of infection. If egg counts exceed 1,000 per gram, then an anthelmintic is typically recommended.

There are three major anthelmintics approved for use in sheep: levamisole, thiabendazole, and ivermectin. There are several other wormer medication products available. Many products can be administered by injection, pills, dry medication mixed with feed, or a liquid drench. Be aware that these products cannot be used for organic production and that you should only use them according to the recommendations provided and under the supervision of your local veterinarian.

An alternative to anthelmintics is to rotate pastures so that sheep do not remain on the same pasture for more than twenty-one days to help reduce infestation. Biological control can be helped by utilizing other species, such as cattle or horses, to graze after sheep to reduce the larval population on the pasture.

Common external parasites include sheep ticks, wool maggots, and lice. A sheep tick, also known as a sheep ked, is not a true tick but rather a wingless parasite fly. It passes its whole life cycle on the body of the sheep, which is typically a nineteen-day period from pupae to maturity. They roam all over the sheep, puncturing the skin and sucking blood to obtain their food. This

Good foot management starts with healthy sheep and by not introducing any sheep into your farm that have obvious signs of foot rot. Examine the feet of each animal you consider purchasing and reject those with redness between the toes, swollen feet, or feet that give off an odor. Routine hoof trimming with hoof clippers will keep feet in correct shape and reduce mobility problems.

There are three main injection sites on the body of sheep that can be used to effectively administer medications while not damaging the valuable muscle portions used for meat cuts. The best intramuscular (into the muscle) injection site is in the neck area between the head and shoulders. Subcutaneous (under the skin but not into the muscle) injection sites are located in areas behind the shoulder, elbows, and rear flanks.

damages the skin and causes irritation and itching as the sheep bite or scratch to relieve those symptoms. Anemia may result in severe cases. There are products available that can eliminate this parasite.

Wool maggots are part of the life cycle of the blow-fly, which are about twice the size of the common house fly. They appear in the spring and reproduce by laying their eggs in masses at the edge of a wound or in manure. One reason for banding the tail of a lamb is to reduce the risk of eggs being laid on their bodies.

There are two different classes of lice that affect sheep: biting and sucking. In either case, the eggs are attached to the individual wool fibers and hatch in one or two weeks into the nymph stage. They emerge as adults after two or three weeks. Lice cause irritation and itching, which can result in restlessness and constant scratching and rubbing against walls and fences. These behaviors can result in lower gains and damage to the wool. Lice are susceptible to commonly used insecticides. You should consult your veterinarian to develop a plan for appropriate parasite control for your area.

VACCINATIONS

Vaccines are used to prevent diseases from afflicting your flock and have very little, if any, effect in treating the disease. These immunizations contain proteins that stimulate the sheep's immune system to produce protection against a particular disease. For example, lambs should be vaccinated for tetanus and enterotoxemia, along with annual boosters. There are a number of vaccinations that can be given at different ages. You should discuss these with your local veterinarian and develop a routine vaccination program for your farm. Some vaccines can only be administered by veterinarians.

The vaccines you need to use on your flock may depend in large part on where you live, your climate, the type of sheep enterprise you have, the purchase of new animals, and the prevalence of nearby sheep flocks.

COMMON SHEEP DISEASES

While no disease can be considered common, there are several that you should be aware of as they are more prevalent than others and have the ability to dramatically reduce the profitability of your flock.

White muscle disease is a problem in many areas. It is caused by a lack of selenium and/or vitamin E in the diet of ewes and/or lambs. It affects the heart and skeletal muscles. The characteristics of white muscle disease are stiffness in the hind legs and a hunched or arched back. Signs of this disease are lambs born dead or lambs that are unable to rise or walk. If they do walk, they do so stiffly. If not treated, it can cause death of a lamb in one to three days but may also affect six- to eight-week-old lambs, often the fastest-gaining lambs. White muscle disease can be prevented by having adequate selenium in the ewe's diet, usually by feeding salt with selenium added. A selenium injection product given subcutaneously at birth is effective if your veterinarian recommends this for your area.

Enterotoxemia is sometimes referred to as overeating disease and is caused by rapidly increasing the energy level of feed. The only symptom of this disease is usually sudden death in an otherwise healthy lamb. However, it can kill sheep of all ages if they consume high levels of carbohydrates. This disease can be prevented by a series of vaccinations and monitoring the energy levels of the feeds in gradual increments.

Mastitis is an infection of the mammary glands and reduces the milk production capacity of the ewe. It can range from mild to severe infections. It is more prevalent at weaning when milk is no longer removed from the udder. The easiest way to prevent mastitis is to halt milk production as quickly as possible by reducing feed and water intake or feeding a very low quality roughage diet with no grain supplementation.

Scrapie is the biggest problem in today's sheep industry. It is a disease of the central nervous system that progresses over a period of years from incoordination to convulsions and finally death. There is no known cure and it is typically spread during lambing season when lambs come into contact with infected placentas, although its devastating effects do not show up until the ewe is three to five years old. Scrapie is a disease of regulatory concern because it is a member of the transmissible spongiform encephalopathies (TGEs), which include chronic wasting disease in deer and elk, bovine spongiform encephalopathy (commonly called mad cow disease), and Creutzfeldt-Jacob's Disease in humans. Producers of breeding stock are encouraged to enroll in the voluntary scrapie flock certification program, which, after five years of scrapie-free monitoring, enables a flock to be certified as "scrapie-free." Sheep can be genetically tested for scrapie resistance. Although it is not a genetic disease, a sheep's genetic make-up influences its suscep-

tibility to scrapie if exposed to an infective agent. If you purchase sheep, it is worth the effort to buy those that come from a flock that is certified scrapie-free.

Ovine progressive pneumonia (OPP) is a slow-developing viral disease that causes persistent weight loss, difficulty in breathing, and lameness or paralysis. It is transmitted from one animal to another or to lambs that drink colostrum milk from infected ewes. There is no treatment or vaccine at present but OPP can be eliminated from the flock using blood tests and removing positive ewes or removing lambs from ewes prior to nursing. Be aware that not all positive-testing sheep will come down with the disease. However, once the signs appear it is always fatal. This is another disease that can enter your flock with the animals you buy, so you need to be careful when buying your initial flock or replacement animals. You should request proof that the flock has been tested for OPP prior to purchase.

When a ewe dies after lambing you will have an orphan lamb or two. Orphans will need a milk replacement product from a bottle or bottling pail. Lambs reared on milk replacer can be successfully weaned from milk feeding at twenty-five to thirty pounds or when they are thirty to forty-five days old. Abrupt weaning is preferable to feeding a diluted milk replacer the last week. Orphan lambs should not be weaned unless they are drinking water or eating feed.

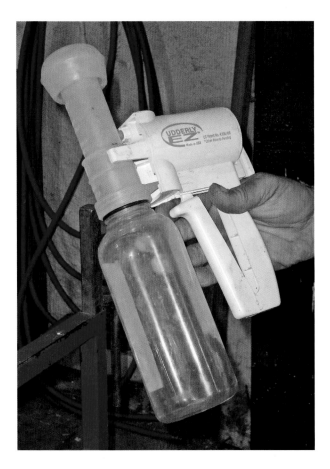

A colostrum extractor is a simple device to use for milking a ewe whose lamb may be too weak to seek nourishment on its own. One teat is inserted in the top of the extractor. Pumping the handle creates a vacuum that draws the milk from the udder. The colostrum can then be fed to the lamb with a nipple bottle.

Pneumonia is a respiratory disease and usually occurs due to stress, changes in weather, poor ventilation, and infectious agents all occurring at similar times. Affected sheep typically become depressed and refuse to eat. They may cough or have breathing difficulties and temperatures are usually over 104 degrees Fahrenheit. You or your veterinarian will need to respond quickly to this condition because it can become acute in a very short time and lead to death.

Coccidiosis is caused by single-cell protozoans that are natural inhabitants of the sheep's intestinal system. It usually occurs in lambs about four weeks of age, when they are susceptible to stress, particularly at weaning

time. Coccidiosis damages the lining of the small intestine and will affect the absorption of nutrients. Diarrhea is the most noticeable symptom and it will appear blackish, bloody, or smeared with mucous. Treatments are available but coccidiosis is mainly a management-related disease typically caused by overcrowding and poor sanitation.

Ketosis is also known as pregnancy toxemia, twin lamb disease, or hypoglycemia and is a metabolic disease that affects ewes during late gestation. It is caused by an inadequate intake of energy during late pregnancy and most commonly affects ewes that are thin, overly fat, older, or carrying multiple fetuses. Ketosis can be both treated and prevented. It is treated by intravenously administering glucose or propylene glycol. It can be prevented by providing adequate energy to ewes during late gestation.

Footrot is caused by bacteria that live in the soil. It is a difficult disease to control once the bacteria have become established on your farm. You can minimize the effects of footrot by maintaining a closed flock. Although the bacteria can only live on top of the soil for up to ten days, once it has attached itself to the feet of the sheep, the bacteria live only in the oxygen-free environment of the hoof tissue where little or no oxygen is present. Footrot starts with a reddening of the skin between the claws of the hoof and if untreated, it progresses to destroy the soft tissue of the hoof. This can cause lameness or misshapen feet and can be easily identified by the foul odor. Routine hoof trimming and hoof bath products such as zinc sulfate can minimize the effects, and vaccines are now available. Check all the feet of new sheep you purchase to determine whether there are signs of footrot present and avoid a flock that exhibits them.

Fungal diseases such as ringworm and pinkeye are highly contagious. They can affect many members of your flock if left untreated. Ringworm is especially a concern with show lambs because they cannot be exhibited in competition while showing signs of this disease. The fungus invades the skin and wool follicles and the fungal spores can be transmitted by contaminated clippers, blankets, combs, bedding, and pens. Lesions are the outward signs of the fungus and can appear anywhere on the animal but most often appear on the head, neck, and back. One of the oddities of this fungus is that it may lie dormant for several years and seem to be gone until conditions become right and another outbreak occurs. Treatments such as iodine and glycerin can be used and applied to affected areas. The fungus often runs its course without assistance from any treatment.

Pinkeye is the common name for the reddening of a sheep's eye caused by bacteria. If left untreated it can affect the eyes and possibly render the animal blind. This condition typically occurs during the warm months and appears to be transmitted from one sheep to another by flies that congregate around their faces and by sheep rubbing their faces together. Pinkeye can be treated with powders, ointments, applications of antibiotics, or the use of homeopathic remedies.

It is very important to understand that fungal diseases and others are highly contagious and easily transferred from animals to humans. Precautions must be taken to ensure that you or members of your family do not come in contact with infected areas. If you treat a sheep with pinkeye or ringworm, be sure to use good hygiene during and after treatment to avoid contaminating yourself. Using disposable latex gloves when treating the infection and cleaning out pens where infected animals live will help prevent the spread of the infection.

Bloat is not a disease but rather the result of rumen gas not being able to escape at a proper rate. The accumulation of excess gas causes the rumen to expand and typically pushes out the left side of the animal. This is a medical emergency and you must assist the affected sheep quickly or it can lead to sudden death. Bloat is more common in lush pastures made mainly of alfalfa and clover when sheep gorge themselves. Giving them dry hay prior to moving from one pasture to another will help minimize the potential for bloat. Treatment can include an oral drench of vegetable oil mixed with water. The oil disperses the gas bubbles and allows the normal return to belching.

RECORDKEEPING

Keeping good records of your sheep business will help you make intelligent management decisions and identify potential problems. There are computer programs that simplify flock recordkeeping, or perhaps written records

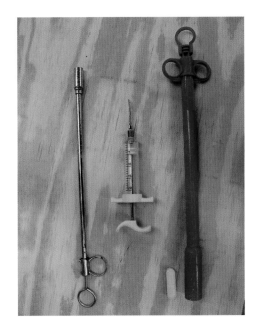

A bolus gun is placed into the animal's mouth and past the tongue. By pushing the plunger, the bolus passes into the sheep's throat and is swallowed.

may be adequate for your needs. As the size of your flock increases, so too does the need for accurate records.

The range of records you keep may include breeding records, lambing records, vaccination records, treatment records, sheep identification records, and anything else that you consider useful. Ear tags are the most easily used recordkeeping tool and they can be attached to the ears of lambs soon after birth when the creation of their record should start.

Your records will most often be used in your management program, but there may be instances when they are required by government agencies if outbreaks of highly contagious diseases occur. Having detailed records available may help with subsequent investigations. If this is not a significant issue, you will have peace of mind knowing that your records are as accurate and complete as possible.

You will become a good manager and develop good husbandry skills with practice and attention to details. Subscribing to breed publications and general farm newspapers will provide you with the latest information on many subjects relating to your farm and sheep-raising program.

LAMBS
AND
LAMBING

Sheep-raising programs are directed toward producing as many live and healthy lambs as possible each year. There are several different systems you can incorporate to increase your flock numbers, regardless of the breed you have chosen. Once you have selected the breed you wish to work with, the next step is deciding what type of lambing system you will use. If you only have a few sheep, it may not be of great concern which system you use since it is likely your lambs will be born in the early part of the year and follow a more traditional lambing schedule. If you decide to raise a significant number of ewes, perhaps fifty or more, there are several options to consider. Regardless of which lambing system you decide to use, the success of lamb survival will be greatly impacted by the care the ewes receive prior to giving birth. Careful management of ewes during the last six weeks of pregnancy will influence lamb birth weights and survivability, and profitability for you.

LAMBING SYSTEMS

After you have selected a breed of sheep, your next decision will be when to have your ewes lamb. There are three components to any lambing system: when lambing occurs during the year, how often a ewe lambs, and whether you use indoor or outdoor lambing. You will have to assess your goals, objectives, marketing strategies, and farming situation to determine which system will work best. Whether you lamb in spring, summer, or fall, there are advantages and disadvantages associated with lambing at different times of the year.

Early lambing is considered to occur from November to February and has several advantages, including labor

Lambing is the most important time of the year for your sheep program because these will be your future replacements or market animals. A successful lambing program is one that minimizes its losses.

The ewe's belly will deepen during late gestation and the udder will increase in size as lambing time approaches. Seventy percent of fetal growth occurs during the last four to six weeks of pregnancy.

availability and marketing opportunities. Lambs born early in the year can be weaned at two to three months of age, placed on a high-grain diet, and fed to reach a market weight of 100 to 125 pounds at three and a half to six months of age at a time when lamb prices have historically been at their peak. The highest prices for lambs tend to occur during the first half of the year, especially around Easter and other ethnic and religious observances. Because of these market demands, winter-born lambs usually sell for higher prices than spring-born lambs.

Lambing in winter allows sheep growers to carry more ewes on their pastures because the ewes are not competing with growing lambs for pasture and use most of their feed for maintenance instead of growth. One disadvantage of winter lambing is the need for adequate facilities during the winter months. Health problems are more likely to occur during winter lambing because of the stress conditions and close confinement. Good management must be applied to minimize the effects of mastitis, scours, and pneumonia.

Late lambing or spring lambing occurs from March to May and is popular because it has several advantages over early lambing, including more efficient use of

forages that come into full production in spring. Lambs go to market in late summer, fall, and early winter as feeder lambs weighing sixty to eighty-five pounds or as slaughter lambs. While the primary benefit to late lambing is lower production costs, it also takes advantage of the natural breeding cycles of sheep. Ewes are most fertile in the fall and many will conceive during their first or second heat cycle, resulting in a narrow thirty-five-day lambing period. With spring lambing, maximum usage is made of spring, summer, and fall pastures when lambs tend to exhibit better weight gains. Pastures must be well managed to take advantage of this system, and predators may be an issue in some areas because of increased feeding requirements of the predator young. Another disadvantage is that lamb prices tend to be lowest in the fall when spring-born lambs are marketed.

Fall lambing occurs from September to November and reduces the need for covered facilities. It has several advantages over the other two systems because it can efficiently use the high-quality forage growth and provides more time to finish slaughter lambs for the traditionally higher winter and spring markets. Weather conditions are generally better for lambing on pastures and there tends

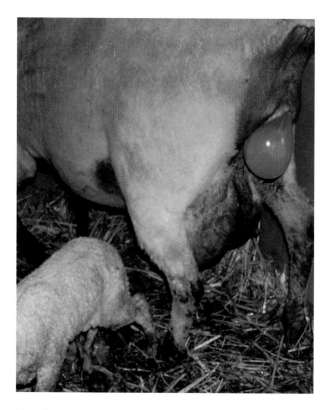

The placenta and fluid, often referred to as the water bag, will appear and then break as contractions increase. The lamb will be born shortly after the water bag breaks if no presentation problems occur. *Jan and Coby Schilder*

to be fewer problems with parasites and predators in the fall. Fall lambing has challenges with conception rates that tend to be lower than with spring breeding. Breeding cycles can be manipulated through light control or hormonal injections to restructure your lambing program.

ACCELERATED LAMBING

Accelerated lambing is one method you can use to potentially increase your lamb crop each year. Accelerated lambing is when ewes lamb more than once a year. The purpose behind an accelerated lambing system is to increase profitability by reducing fixed costs and producing a more uniform supply of lambs throughout the year from your flock. There are several systems that you can consider for use in your program. To be successful with any of them, your management will greatly influence the outcome. The first accelerated lambing system, twice-a-year lambing, is the most intensive form when your ewe produces two

lamb crops a year. Although it will potentially maximize your lamb production, it is difficult to achieve uniform results under most commercial situations.

Another form of accelerated lambing is keeping your ram(s) with the flock on a continuous basis so they can mate with the ewes whenever any come into heat. This will scatter the lambing across a longer period of time but will make it difficult to know when lambs are due. This will also influence the timing of vaccinations and deworming, and providing supplemental feeds to an unbalanced group of lambs may make it more difficult.

Three lamb crops in two years is the most common system used for accelerating lambing and typically results in an average lambing interval of eight months. This creates an average of 1.5 lambs per ewe per year and a 40 percent increase in production has been reportedly achieved. This system follows a schedule such as April mating and September lambing, December mating and May lambing, and August mating and January lambing, or other variations.

The STAR system is another accelerated lambing program developed by Cornell University. The goal is for ewes to produce five lamb crops in three years. The calendar year is divided into five segments, like the points of a star from which its name is derived, that represent one-fifth of a year, or 73 days. Two-fifths of a year is 146 days, the approximate gestation length of a ewe. There are five lambing periods each year and the flock is divided into three groups and each is managed separately. The first group consists of breeding and pregnant ewes and rams. The second contains lambing and lactating ewes and lambs. The third group is the growing lambs, which may be market lambs or replacements. If a ewe misses a breeding, she can still lamb three times in two years. Another advantage of the STAR system is that it is natural and does not use hormones or light control to achieve out-of-season breeding. However, it does require that you select sheep that breed during any season.

Whether you use an accelerated lambing program or not may depend on several factors, such as providing more intensive daily management, keeping accurate records, compensating for the additional costs and labor by increased sales and income, and your time commitment to establishing a viable and successful program.

EWE GESTATION

A fetal lamb will gain 70 percent of its growth during the last four to six weeks of pregnancy. As the fetus grows, the ewe's rumen size decreases due to lack of room inside the body cavity. This results in a greater need for nutrients, especially if the ewe is carrying twins or triplets as the growing fetuses require additional nutrition as well. Providing an adequate diet for pregnant ewes will prevent ketosis, or pregnancy toxemia, and will support the development of the mammary tissue for lactation. Adequate nutrition during the late-gestation period will ensure that lambs are born with appropriate weights to aid survival and that ewes will be able to produce a sufficient milk supply. Failure to provide adequate nutrition during this period can lead to small or weak lambs, higher lamb mortality, poor milk yield and colostrum quality, and possible pregnancy toxemia. While it is not desirable to underfeed a ewe during this period, it is also undesirable to overfeed her. Overfed ewes can become fat and are more prone to ketosis and can experience

As the head and shoulders of the lamb begin to appear, the umbilical cord is torn away from the inside of the uterus. The lamb is now in a critical position as it must begin breathing on its own. *Jan and Coby Schilder*

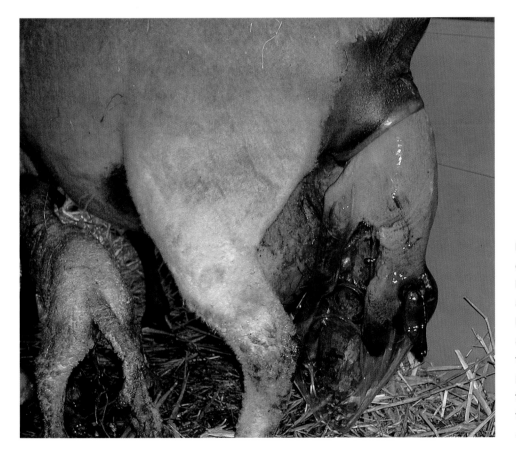

It is not unusual for the ewe to be restless during lambing. She may get up and down several times before the lamb is born and may even stand while trying to expel it. The placenta must be cleared from the lamb's face so that it does not suffocate. *Jan and Coby Schilder*

The first life indications from a newly born lamb are open eyes and attempts to breathe. You may need to clear its mouth or hold it upside down to drain fluid that may still be lodged in its throat. Rubbing the lamb will help initiate breathing if it does not appear it is doing so. *Jan and Coby Schilder*

more difficult births. Overfeeding can also result in large fetuses that make birthing difficult.

Providing essential nutrients and vitamins is important during the gestation period. Selenium and vitamin E are critical nutrients to be supplied in adequate levels. Selenium is passed from the placenta to the fetus, and low levels have been associated with poor reproductive performance and retained placentas after giving birth. Adequate selenium intake can also prevent white muscle disease.

Milk fever is caused by low blood calcium and can affect lactating ewes when there is an inadequate intake of calcium. During late gestation, a ewe's requirement for calcium almost doubles, so the addition of calcium into the diet is needed. Just as too little calcium can be a problem, too much calcium can cause a condition where the ewe fails to mobilize her reserve supplies, resulting in her inability to stand or walk. An intravenous calcium treatment must be given to quickly correct this deficiency or the ewe may die.

It is easier to correct a calcium deficiency or prevent overfeeding of calcium. Grains such as corn, barley, wheat, and oats are poor sources of calcium. Legumes,

In the case of multiple births, the ewe will attend to the lambs when the final lamb is born. Licking and talking with her lambs helps bond mother to offspring as they respond to her touch. *Jan and Coby Schilder*

Lambs are on their feet and looking for something to eat minutes after birth. The first few days are a very critical period in their survival. *Jan and Coby Schilder*

including alfalfa and clover, are good sources and should be reserved for feeding during the lactation period. Grasses and mixed legumes are typically lower in calcium and are best fed during late gestation. If the forages are low in calcium, they can be supplemented by a complete grain mix or mineral supplements specifically formulated for sheep. There is a delicate balance between having enough stored calcium in a ewe's body to meet her needs once she lambs and having too much. After lambing, the ewe's system quickly requires calcium to meet her milk-production needs. This dramatic decrease in available calcium levels is the cause of milk fever, which needs to be corrected to keep the ewe alive. Attention to her feeding program will help ensure that she survives the lambing process.

VACCINATIONS AND DEWORMING

If you are following an organic production program, you have fewer options available for vaccinating and treating your pregnant ewes than if you produce for a nonorganic market. While vaccines are allowed under organic rules,

parasiticides are generally not allowed for regular use. Products such as ivermectin may be used in emergency cases and only when the animal is not producing organic milk. Parasites can be controlled through pasture management, allowed herbal remedies, and breeding for resistance. No antibiotics or hormones are allowed in organic production. However, an organic farmer cannot withhold medical treatment to preserve the organic status of an animal. If antibiotics are used as a last-resort remedy, the treated animal cannot be considered for further organic production.

If you pursue a nonorganic production system, your options for treatments are greater. Pregnant ewes should be vaccinated for clostridial diseases three to four weeks prior to lambing. Vaccinated ewes will pass these antibodies to their newborn lambs through the colostrum milk. If you do not know the vaccination status of your ewes, you can provide protection by vaccinating them twice at least two weeks apart. At this time it would be good to include your ram(s) in the program.

Deworming ewes prior to lambing is an important consideration as internal worm eggs will grow once a

Colostrum is the first milk that a ewe produces after lambing. It has high levels of several nutrients important for lamb health. It also contains a high level of antibodies against a variety of infectious agents. Lambs should nurse during their first thirty to sixty minutes of life.

ewe becomes pregnant and she has a loss of immunity because of hormonal changes. Deworming will help the ewes expel the worms and decrease the exposure of newborn lambs to worm larvae. Ridding the ewes of worms during the late-gestation period will also decrease the number of worm infestations in the pastures once they are outside. Study the products that are available if you decide to use a deworming protocol because some should not be used during certain periods of the pregnancy.

You should also consider feeding a coccidiostat during late gestation. Coccidia are bacteria present in all sheep digestive systems that can cause fevers and heart ailments if affected by high concentrations. Feeding a coccidiostat will reduce the number being shed into the lambing environment.

SHEARING

Ewes should be completely sheared or at least crotched about a month prior to lambing. While shearing removes all the wool on the ewe, crotching removes just the wool around the udder and vulva. Removing the wool around the udder allows the newborn lambs easier access to the milk supply and provides a drier environment because shorn ewes produce less moisture. Shearing also makes ewes less susceptible to lying down on their lambs because their bodies are less cumbersome. Removing the wool around the vulva will provide a cleaner area for birthing and for any manual assistance that might be needed to assist those births. If shearing occurs during a cold part of the year, you will have to provide more nutrition to compensate for the ewe's body heat loss. Shorn ewes will also require adequate shelter but will typically take up less space around feeders or in pen areas.

FACILITIES

You should provide a clean, dry area for your ewes to lamb. Pens used for birthing should be well-bedded with straw, wood shavings, corn fodder, or other dry materials. Drafts should be eliminated at floor level because these will affect the newborn lambs.

A lambing pen is called a jug and should be ready before the first ewe lambs. Providing a space twelve to fourteen square feet per ewe should be sufficient unless you have a breed with very large ewes. You should have enough pen space to house 10 percent of your flock at one time. Additional pens may be needed if your lambing program is more concentrated. A pen size of four feet by four feet should be adequate for small ewes; larger ewes may need something a little bigger. Depending on the buildings available on your farm, you may be able to construct different-sized pens to suit your needs.

Shed lambing is the most common system of lambing. Many producers who raise their sheep on pasture or range usually bring their ewes in for lambing.

Shed lambing provides protection from the elements and predators. It also allows for earlier lambing and different lambing systems such as accelerated lambing. You have more control in shed lambing to assist with problems any ewes may have during the birthing.

Hypothermia and starvation are the two principle causes of lamb death. Creating a warm environment and getting colostrum into a newborn lamb will help raise its core body temperature. Drying lambs off by using towels or putting them in a warming box, shown here, are ways to elevate their body temperatures. A hair dryer inserted into the box provides warmth.

Pasture lambing is increasing in usage due to an emphasis on more sustainable systems of livestock production that require less input, including during labor. Pasture lambing is considered more natural but must occur during periods of mild weather to prevent significant lamb losses. Predators can be a problem at lambing time and attacks may become frequent unless livestock guardians are employed. You may have a more difficult time assisting any ewe that may have problems during lambing while on pasture. By first assessing your available buildings, goals, expertise at handling problems, and lambing system to be used, you will be in a better position to determine whether you will likely be more successful with shed or pasture lambing.

PREPARING FOR LAMBING

Once you have provided a clean, dry pen for your ewe(s), there are several things you can have ready for the time

the lambs begin to arrive, including heat lamps and a hospital kit. Providing heat lamps for the newborn lambs will help them acclimate to their new environment, particularly if they are born during cold weather. While they may be sheltered, young lambs can benefit from the warmth provided by lamps until they gain enough strength to no longer need them.

A 250-watt infrared heat lamp should be satisfactory, but be sure it is safely mounted and secured and that the ewes cannot dislodge it. A dropped heat lamp can create a fire hazard if it comes in contact with dry bedding. Providing a separate area with the heat lamp into which the lambs can retreat after nursing will allow them to stay warm when away from their mother. Lambs should not have access to heat lamps for an extended period after birth. Three or four days under a heat lamp should be sufficient to dry them completely unless it is extremely cold or the lambs don't appear to be progressing. Additional warmth may relieve some of the stress leading to illness. Having a hospital kit on hand and stocked with birthing aids, soaps, towels, and other items will reduce the need to search for them when time may be short to save a lamb.

SIGNS OF LAMBING

There will be several physical changes in the ewe to indicate when lambing time is imminent. The most noticeable is that the udder will begin to swell and extend, which usually happens several days prior to giving birth. The swelling will increase as the time gets closer to lambing and the vulva will become enlarged and have a soft, shiny appearance. The ewe may not want to eat as the time approaches and may appear restless by getting up and down numerous times. She may prefer to be alone. If the lambing occurs outdoors, she may go off by herself to lamb. All of these characteristics of lambing are normal. This is a natural process that has evolved over thousands of years and most ewes will give birth to their young without any problems. However, having an understanding of the lambing process will help you to assist ewes if it becomes necessary.

There are three distinct stages in lambing: the dilation of the cervix and opening of the birth canal, the appearance and expulsion of the fetus, and the discharge of the placenta that signals the end to the entire process. A hormonal change occurs during the final stages of the pregnancy when the progesterone levels that maintained the pregnancy begin to fall and signal that it is time for birth. This process may last for only three to four hours or as long as twelve to twenty-four hours. The contractions in the uterus coincide with the opening of the cervix and a clear-whitish discharge will appear. This discharge is the sealing material that lined the cervix and kept contaminants from entering the uterus during the pregnancy. Once this is expelled, the contractions will increase and begin to push the placenta until it breaks and releases the fluid that has surrounded the growing fetus. This is typically referred to as the water bag and once broken, the birth of the lamb should occur within an hour or less. If nothing appears to happen within that time, you may need to provide assistance.

The tip of the nose and front feet of the lamb should appear first during a normal presentation. Once expelled, the lamb will begin to breathe and the mother will quickly rise to investigate her new offspring and begin to lick it dry. This bonding may not last long in the case of twins or triplets and her attention will be quickly drawn from her newborn to deliver another lamb and the birthing process is repeated.

The final stage of the lambing process is the expulsion of the placenta. This was the lining of the uterus through which the nutrients were transferred from the mother to the fetus. The release of oxytocin triggers two actions: the expulsion of the placenta and the initiation of lactation. The placenta is usually expelled thirty minutes to two hours later. If it is retained longer than twenty-four hours, it could indicate a problem for which you may want to get veterinary assistance. Be aware that with multiple births there will also be multiple placentas: one for each lamb born. You should remove the expelled placenta(s) from the pen as the ewe will often try to eat it. This is an ancient instinct that tells her to hide evidence of lambing from predators to protect her young. Eating it will not cause any health problems, however, as it is quickly digested. Problems may arise if she consumes two or three placentas in a short time, so it is best if they are removed. She won't notice that they are gone as she will direct her attention to her newborn.

Lambs born in early spring can be subjected to extreme weather conditions after they are born. It is important to provide them with a clean, dry area where they can bond with their mothers.

SIGNS OF LAMBING
+ Udder will begin to fill
+ Ewe will stop eating a few hours before delivery
+ Ewes go off by themselves
+ Restlessness, raising their tail, stretching, lying down with the head extended and obvious straining
+ Strong ewes will get up and down immediately preceding and often during delivery

DIFFICULT BIRTHS
If the lamb is not born within about one hour after the water bag has broken, it may need assistance. There are several abnormal presentations that can make a birth difficult and will require your assistance or the assistance of your veterinarian or another experienced person. You can become knowledgeable about assisting with difficult births to reduce the time waiting for help to arrive. Veterinarians are usually busy with many calls during their day and may not be able to arrive quickly. The distances involved may prevent a quick response to your farm. With diagrammatic aids to help visualize what is occurring within the ewe and with a little hands-on experience, you can become adept at correcting these problems and assisting your ewes through a difficult birth.

There are multiple abnormal lamb positions, but the six most common ones involve backward or breech presentations, head back, legs back, twins or multiples, elbow locks, and upside down. There are other problems that do not necessarily have to do with positioning but are abnormal or pose difficulties during delivery. These include lambs of disproportionate size that cause tight births; ringwomb, which is a failure of the cervix to dilate or open; swollen heads, a condition arising from a lamb's head being outside the vulva for a long time without the rest of its body being expelled; and dead or deformed lambs.

PROVIDING ASSISTANCE

A normal lamb delivery is when the two front feet appear with the head resting between them, and assistance rarely will be needed. In cases where the lamb is too large for a small ewe to deliver, you may have to assist.

Knowing when to assist a ewe during lambing is a difficult thing to recommend because each birth is unique, although there are many similarities between them. If a ewe has been straining for an hour or more, you should begin to closely assess the situation. This can be done by reaching into the birth canal to determine the nature of the problem. Before you do this, wash your hands and arms with warm, soapy water and use a non-irritating lubricant to coat your hand and forearm. Make certain that you do not have sharp fingernails as these may cause injury to the inner lining of the uterus. Use the soapy water to cleanse the vulva to make certain no manure or foreign materials will enter the ewe when you insert your hand.

You can make the entry into the birth canal easier by bunching your fingers together in a forward cone shape and tucking your thumb in while you slowly insert your hand. If your ewe is still lying down, the pressure of her straining and the presence of the lamb may make entry with your hand a challenge. By slowly pushing back against the lamb you should be able to determine the nature of the problem. If it is a foot or leg problem, you may have to push the lamb back far enough to get the leg pulled forward.

If you need to push the lamb(s) back into the ewe's body cavity to free a leg or a head, be careful not to punc-

An artificial heat source may be needed to help lambs maintain body heat during extremely cold weather.
Using heat lamps is one means of providing warmth for individual lambs or groups.

ture the uterus while repositioning the lamb's hooves. Sometimes the birthing process has progressed to the point where the pressure against your hand is too great to achieve any success in repositioning the lamb. If this occurs it might be wise to try to get the ewe to her feet so that the lamb will settle down into the body cavity and allow you to reach in more easily. At this point, the ewe may decide she doesn't want to get up and you will have to do the best you can. As uncomfortable as it may be, sometimes it is better to sacrifice the lamb for the safety and survival of the ewe.

There will be times when you provide all possible assistance but are still not able to save the lamb. This should not be taken as a defeat but rather as a challenge to do better or determine if you could have done anything differently. It is a rare birth where you cannot learn something from the experience. Saving the ewe will allow you to return to your breeding program faster than losing her and raising the lamb as an orphan.

A live lamb will often help with its own birth once the repositioning has been achieved. Try to avoid excessive pulling and straining as this can harm or even kill the lamb. If you do need to pull on the lamb, make certain that the ewe's cervix is completely dilated. Trying to pull a lamb through a partially dilated cervix will be difficult and can seriously damage the uterus by tearing it.

DIFFICULT BIRTH ASSISTANCE

Helping deliver a lamb requires cleanliness, patience, perseverance, and gentleness. Rushing a birth does not work well and can lead to further complications and permanent injury to ewes. When lambs need to be pulled, a steady, gentle but firm pressure works best. Some birthing tips include:

- Wash your hands and arms with soap and water.
- Cover your hands and the ewe's vulva with a disinfectant liquid soap.
- Apply lubrication to your hand and arm.
- Be sure your fingernails are closely trimmed before assisting or use latex gloves if available. Avoid tearing the cervix with your fingernails, manually stretching the cervix, and pulling too rapidly.
- Never apply traction to the lamb until it is in a position to be properly delivered.
- Assemble supplies prior to lambing that can assist in pulling lambs. These supplies include lubricant, head snare, restraining halters for ewes, antibiotics for the ewe to minimize infections, towels for rubbing lambs, tincture of iodine for disinfecting navels, lamb feeding tube for weak lambs, and lamb reviver.
- Intrauterine antimicrobial boluses should always be placed deeply in the uterus after birth if you have entered the birth canal with your hand or arm.
- Women of child-bearing age or pregnant women should avoid working with pregnant ewes or be especially cautious during lambing. There are a number of abortive-type diseases that can be transmitted from sheep to women.

- If you feel too inexperienced to handle difficult births, have someone with experience on hand at lambing time so that you can learn.

ABNORMAL PRESENTATIONS

Backward or breech lambs typically cannot be turned around to enable a forward birth, but this type of birth can still be considered a normal delivery. Backward or breech births are more common in the case of twins or triplets. The simplest way to extract the lamb is to pull it out in its backward position. It needs to be done quickly once the umbilical cord is torn. As the lamb passes through the birth canal the lamb will automatically want to begin breathing and may take fluid into its lungs.

If the lamb is breech and you are able to bring it out quickly, it should have the ability to start breathing if you hold it in the air by its hind legs and allow any fluid that might have been inhaled to drain out of its mouth. Vigorously rubbing the lamb and its lung areas may awaken an otherwise comatose lamb. In these cases, having an esophageal tube or baby lamb reviver on hand may be very useful. You can also use something as simple as a piece of straw or hay and push it gently into its nose to create a sneezing reaction to take a breath of air. Mouth-to-nose resuscitation may be needed if no other aids are available.

In rare instances, a breech or backward lamb may be upside down. If this should happen, you will need to turn the lamb rightside up or at least on its side before pulling it out backward. If you don't, the angle of the ewe's pelvis will not conform to the spine of the lamb and you may break the lamb's back. During any assisted birth, whether or not there are positioning or leg problems, you should pull the lamb firmly and slightly downward toward the ewe's udder or hocks and not straight out. This places less stress on the ewe's pelvis and the skeletal structure of the lamb.

A head-back presentation requires you to push the lamb back into the ewe's uterus to the point where you can turn the head straight. This usually quickly corrects the problem and allows an easy birth.

There are a number of different legs-back combinations that create difficult births. The number of combinations increases with multiple births. The legs may be back with a breech birth and you will need to push the lamb farther into the body cavity so that you can catch the hind feet and pull them toward you. With a head-forward presentation, one or both front feet may be back. Legs back may occur in one or both of the twins and you will have to sort one out first before attempting to correct the second. With twins or triplets it is sometimes easier to correct the legs-back problem because there is more room to maneuver as the uterus hasn't been able to sufficiently contract. An elbow lock is typically a normal birth but with the lamb's elbows locked in the birth canal. Pushing the lamb back into the canal should allow you to extend the leg for easy delivery.

Keeping good records of difficult births may assist in your management decisions. Ewes that consistently have birthing problems should be considered for culling. These problems would include presenting dead or deformed lambs, which may be symptomatic of a genetic predisposition. Dead lambs will sometimes require veterinary assistance, particularly if it has been dead for some time and has started to decompose. While the uterus is a great incubation machine for a growing lamb, it is also a great incubator for germs and bacteria growth if they enter through the birth canal. A dead lamb that has not been expelled or removed quickly becomes a health threat to the ewe.

If your flock experiences many lambing problems, you may want to reassess your breeding or nutrition programs. Having many oversized lambs may indicate an excessive feeding program during late pregnancy or using a ram that sires lambs that are too large. Making sure that the first lamb ewes have reached an acceptable size prior to breeding may also reduce problems.

MALPRESENTATIONS OF LAMBS

+ *Tight birth:* Caused by a large lamb, a small ewe, a small pelvis, or any combination of the three. Slow, steady pressure with the assistance of lubrication may correct the problem if the head and feet are in a normal position.
+ *One Leg Back:* Head and one front leg come normally while one leg is not extended into the birth canal. Lamb may need to be pushed back into the ewe to pull the leg forward.

Pasture lambs are easier to approach and work with the day they are born and through the second day. That is when the tagging and docking of tails should occur because by the third day of life they are more difficult to catch in the open and you may need to corral them in order to work with them.

+ *Head Only, Two Front Legs Back:* Prompt action is needed for this serious situation because both feet will need to be pulled forward, which will require pushing the lamb back into the ewe's body cavity.
+ *Head Back, Front Legs Only:* The head must be brought into place for the lamb to be born. If siblings are involved, be sure you have the right head before pulling it forward with your hand or head snare.
+ *Backwards, Rear Legs First:* This can be a normal delivery and is common with twins and triplets. It is not necessary to turn the lamb around, but once the lamb is pulled past the umbilical cord you will need to quickly free the lamb from the birth canal or it may drown in the placenta fluids.
+ *Breech:* The lamb is positioned backward with the rear legs tucked under and only the tail appears. You will need to bring the rear legs forward by cupping the feet in your palm. You may need to push the lamb back into the ewe to reach the feet.
+ *Locked Elbows:* This occurs when there is not enough room and the head is forced into the birth canal and

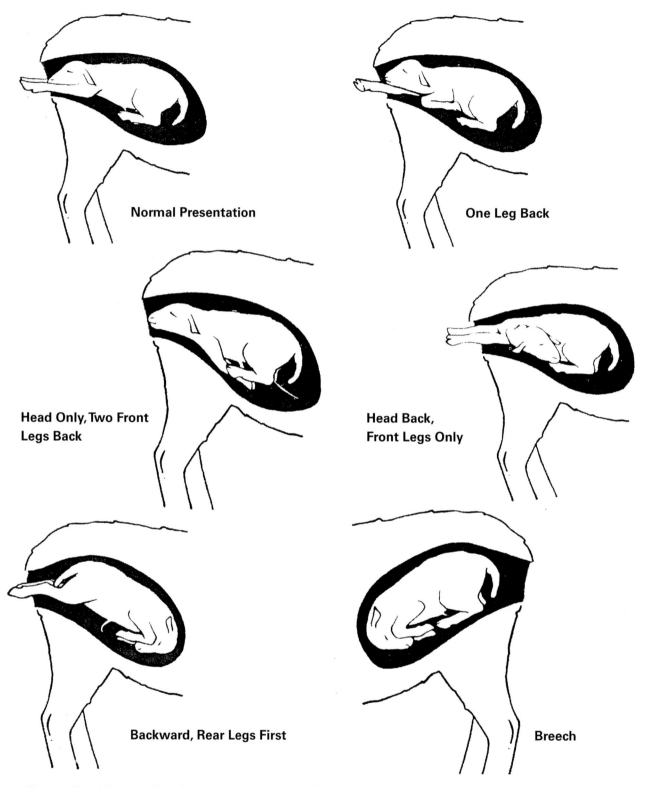

Normal Presentation

One Leg Back

Head Only, Two Front Legs Back

Head Back, Front Legs Only

Backward, Rear Legs First

Breech

Abnormal Lamb Presentations: Ewes may need assistance if they do not make significant birthing progress forty to sixty minutes after the water bag appears. This is a good indication that there may be a malpresentation or a problem between the lamb and the ewe's pelvic size. These diagrams illustrate a number of abnormal presentations at birth and will assist you in

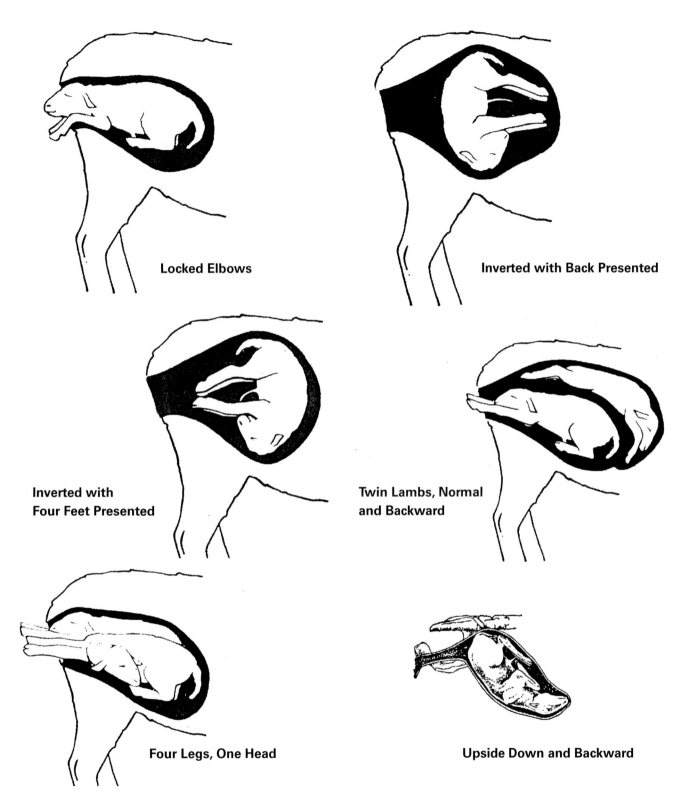

Locked Elbows

Inverted with Back Presented

**Inverted with
Four Feet Presented**

**Twin Lambs, Normal
and Backward**

Four Legs, One Head

Upside Down and Backward

understanding different possible positions. You will need to visualize lamb parts and positions by touch to successfully pull lambs. *Dr. Norman Gates, Washington State University, 1942 USDA Special Report*

the elbows are forced downward. In effect, the elbows are locked behind the ewe's pelvis and the more forceful the ewe's labor, the tighter the lamb becomes locked. Push the head gently back into the canal to relieve the pressure. At the same time with your other hand, pull on one or both front feet until you feel the elbow pop out of its locked position.

+ *Inverted Lamb with Back Presented:* You should either position the lamb for a backward delivery or for a normal two front legs and head delivery. Initial diagnosis may be a challenge when reaching in to locate a body part.

+ *Inverted Lamb with Four Feet Presented:* Identification of the front and rear legs is needed. Once it is determined, you can decide whether to use a front or backward delivery.

+ *Twin Lambs, Normal and Backward:* Deliver the normally positioned lamb before assisting with the backward lamb. Identification of each twin is necessary to ensure enough room to deliver the first lamb.

+ *Four Legs, One Head:* This is a fairly common occurrence with twins and you should attend to it quickly. Both lambs must be gently pushed back into the uterus to allow room to deliver one in a normal position. The second lamb can be manipulated into a normal position as a "head back, front legs only" delivery.

+ *Upside Down and Backward:* This is a difficult birth where the lamb needs to be turned inside the ewe so the rear feet point downward when pulled out. The turning may wrap the umbilical cord around the body or neck of the lamb.

POST-LAMBING

There are a number of considerations once the lambing process has been completed. The removal and disposal of the expelled placentas or afterbirth should be done so the ewes don't eat them. Sometimes that is not possible and they beat you to it. Aborted lambs or lambs born dead should also be removed from the lambing areas and disposed of, typically through burial.

You will want to make sure that the ewe claims her lamb. Moving mother and baby to a separate pen will help with bonding and prevent the ewe from confusing her lamb with another ewe's lamb. Ewes that don't exhibit a bonding behavior with their lambs should be noted and considered for removal from the flock. If a ewe doesn't claim her offspring, her value to the future of your flock is greatly diminished. Bottle-fed lambs, while able to survive and grow, require more labor and have a higher mortality rate than lambs born to ewes with a great mothering instinct.

On rare occasions, a ewe will have a vaginal prolapse. This is when the uterus expels itself and comes out the vulva. In most cases it is not life threatening unless the blood vessels inside the tissue of the uterus become torn as it is suspended from the vagina. A prolapse will generally need veterinary assistance and sutures are used to keep the uterus from coming out again. A process of healing needs to occur to return the uterus to reproductive fitness. Ewes that exhibit a prolapsed uterus should be considered for removal as there is some evidence that this is genetically inherited.

NEWBORN LAMBS

After being cocooned within a floating sea of fluid for five months, the newborn lamb now seeks nutrition and quickly finds the udder and teats of the mother. A nursing lamb is one of nature's wonders and a never-ending source of amazement. Silent signals occur after the lamb is born to let it know where to look for milk. Most are successful in finding the udder, but some may need your assistance by directing them to it.

It is important to get your newborn lambs off to a good start because nearly 20 percent of lambs die before weaning, and 80 percent of those deaths occur during the first ten days. Decreasing your lamb losses to a more acceptable 4 percent begins with making sure they receive an adequate supply of milk. The first milk produced by a ewe after giving birth is called colostrum and contains a high level of nutrients important to the health of the lamb. Colostrum contains antibodies used to fight a variety of infectious agents. Newborn lambs do not carry any of these antibodies because they do not cross the placenta from the ewe's bloodstream. The greatest absorption of the antibodies by the lamb occurs within the first twenty-four hours. After this time period, the

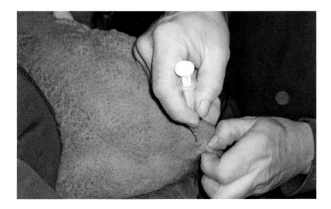

Tail docking is a simple, painless procedure when an anesthetic injection is made into the tail area. Once the area is numbed, a rubber band is placed near the top part of the tail. It is good management to vaccinate the dams against tetanus if you are docking tails as this will give the lamb some immunity against infection after it is born.

ability to absorb antibodies decreases until it becomes negligible. All lambs need colostrum, although it is possible for them to survive without it. However, the likelihood of lambs dying from disease is higher in lambs that do not receive it.

Monitor your newborn lambs to make sure they nurse. It is recommended that lambs receive about 10 percent of their body weight in colostrum during the first twenty-four hours. This means a ten-pound lamb should consume one pound of colostrum. While it may be impossible to determine if the lamb is actually consuming that amount, you can ensure that it will get some amount of colostrum by checking the teat openings on the ewe's udder. You can hand strip each teat to remove the wax plug to make sure the end is open and a stream of colostrum is available to the lamb. Some lambs may not begin to nurse and may need assistance. Colostrum powders are available that can be mixed with water as a substitute for its mother's milk. Lambs that do not nurse will have to be bottled or tube fed with colostrum to get enough nutrition. In many cases, cow colostrum can be used if a ewe's colostrum is not available.

Disinfecting navel cords on lambs soon after birth will block a possible route for infectious agents. Long navel cords can be clipped shorter or left alone as they will dry up in several days. Using an iodine solution or other antiseptic to clean the navel cord should be sufficient to prevent infection.

Newborn lambs need to maintain a body temperature of 102 to 103 degrees Fahrenheit to avoid hypothermia or low body temperature. Hypothermia causes many lamb deaths because it lowers the core body temperature of the lamb. A newborn lamb must produce as much heat as it is losing to the environment. This is one reason heat lamps can be helpful in keeping lambs warm. Large lambs can tolerate heat loss better than small lambs, and lambs with thicker coats will lose less heat than those with light coats.

The second principal cause of early lamb mortality is starvation, which is a result of not getting enough nutrition during the first few days. Lack of nutrition may have several causes, including failure to drink colostrum, rejection by the mother, lack of milk production by the ewe, mastitis, teats that are too large or too close to the ground, or the effects of a difficult birth. Starvation is the largest killer of lambs. Lambs with drooping heads and ears or those who are too weak to stand are showing signs of starvation. Closely monitoring your newborn lambs and watching for those that don't exhibit normal behavior when compared to others may help alert you to potential problems before it is too late.

WEANING LAMBS

Weaning occurs when lambs are removed from the milk diet provided by their mothers. Weaning can occur between four weeks and five months of age, depending of a variety of factors. When lambs are born, they cannot digest anything but milk until about three weeks of age. As they grow older and larger, lambs can slowly be converted to a forage or grain-based diet. This will typically coincide with the ewe's milk supply, which reaches its peak at about four weeks and slowly declines by half over the next six weeks. About 75 percent of all milk produced by the ewe during her lactation occurs during the first two months.

Weaning typically occurs at about sixty days and usually prior to ninety days. The size of the lamb is generally more important than its age when considering weaning. A well-grown lamb should weigh about forty-five pounds at sixty days of age and can be successfully

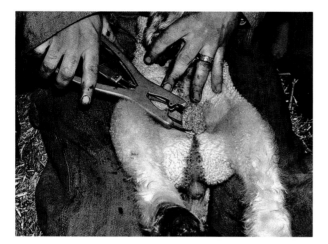

Castration of male lambs can be done through surgery or a more convenient method such as banding. This is a similar process to tail docking where a rubber band is stretched over the scrotum and placed near the body wall. It is a bloodless and less traumatic method than using a surgical knife.
University of Wisconsin Sheep Research Station

An elastrator can be used for tail docking and castration. The small round band is made of stiff rubber that is stretched to place over the tail or scrotum. It is an easy-to-use and effective method of sterilization and docking.

weaned providing it has received a creep feed during that time. Lambs are efficient converters of feed grains and transferring them from milk will help with weight gain. As the lamb's diet changes to forage and grain, the milk from their mother becomes less important and the ewe will increasingly use her nutrition for maintenance rather than milk production and return to breeding condition.

The time of weaning may be affected by outside influences other than lamb weight or age. During times of extreme drought it may be easier to wean them early so that pastures may be used more efficiently. Early weaning may also allow for earlier culling of ewes no longer needed in the flock.

Late weaning usually occurs in the fall when ewes come into heat. In a traditional lambing program this would be at about six months of age. By this time, the ewe's milk production has dropped significantly and weaning will have the effect of producing less mastitis and causing less stress for the ewe and lambs.

Newly weaned lambs should be monitored for health problems and you should be alert to overeating. Lambs tend to be more stressed at weaning time than ewes because they have lost their mothers as well as having to locate their own nutrition. Most lambs do not yet have fully developed immune systems, and during this transition they are more susceptible to diseases. It is important to maintain a constant diet during the first week after weaning, which will help reduce the stress level for the lamb. More information on feeding lambs postweaning is found in Chapter 7.

MASTITIS

Mastitis is the inflammation of the mammary gland due to infectious agents entering the teat canal. While mastitis can occur at any time during lactation, it is more prevalent at the time of weaning because the milk is not being drawn from the udder and the pressure from the remaining milk causes leakage from the teat end through which bacteria can enter and cause infections. Once mastitis occurs in a ewe, her productive capacity is significantly diminished and may render her useless in your flock's future. A ewe that doesn't produce milk will not be able to raise her young. Udder infections destroy the milk-producing tissues and causes scarring, severely limiting the productive capacity of the remaining tissue that may not have been infected. A whole quarter is infected while the others may not be, although these bacteria can be easily transferred when the lamb(s) nurse from one teat to another.

The chance of mastitis occurring can be diminished at the time of weaning through the diet the ewe is fed. Removing grain, high-protein hay, and water, and feeding a low-quality diet will decrease milk production and the chance for mastitis. Care should be taken with decreasing the water amount in times of high temperatures. Sound management practices that will reduce milk production at the time of weaning will help minimize problems.

DOCKING AND CASTRATION

Almost all lambs are born with a tail and its purpose is to protect the anus, vulva, and udder from weather extremes. Today's production systems use tail docking as a health and management practice, and you will have to decide if you will dock your lambs' tails or not. Removing the tail prevents fecal matter from accumulating on the tail and hindquarters, making shearing easier and greatly reducing a harbor for wool maggots, which can attack the skin.

Although about 97 percent of all lambs in the United States are docked, not all sheep need their tails removed. Hair sheep do not have long, wooly tails or wool on the underside of their tails, nor do short-tail breeds developed in Northern Europe, which makes it unnecessary to dock their tails.

The easiest method for docking tails is banding, where a rubber ring is applied to the tail using a device called an elastrator. This handheld pliers stretches the rubber band which allows it to be slipped over the tail; when released, the rubber band fastens itself around the tail. Banding is a bloodless and relatively painless method. Injecting a small dose of a local anesthetic, such as lidocaine, at the top of the lamb's tail near the tailhead will eliminate the pain the lamb experiences. When the painkiller wears off after a short time, the lamb does not know it has been banded. The band stops the flow of blood into the tail and causes it to die. Once the tail is dead, it will generally fall off within ten days of banding. Some states have developed policies mandating minimum tail lengths for lambs shown by youth. Check with your local veterinarian or county agricultural extension agent for the regulations affecting your area.

Castration is the removal of the testicles from male lambs; those males that are castrated are called wethers. Not all male lambs need to be castrated, unless you plan to raise them with females. If male lambs are marketed prior to six months of age, they typically are not discrim-

A shepherd's cane is a useful tool for snaring newly born lambs that otherwise might elude capture for tagging. The hook on the cane is used to catch a hind leg in a humane way that does not cause injury.

inated against in price. Depending on your market, you may prefer not to castrate the males. Some ethnic buyers prefer intact males and may pay a premium for them.

At the time of buying your first sheep, you should decide which management practices you will employ and which you may prefer not to use. Docking and castration are two of these practices that you will have to decide upon. Use your common sense judgment and consider your goals, philosophies, and health concerns for your flock to determine what is best for you and your sheep-raising situation.

Here is a list of some breeds that do not require tail docking:

Hair Sheep	Short-Tailed Breeds
American Blackbelly	East Friesian
Barbados Blackbelly	Finnsheep
Dorper	Icelandic
Katahdin	Karakul
	Romanov
	Shetland

CHAPTER 12

PREDATORS
AND
LIVESTOCK
GUARDIANS

Keeping sheep safe from predators is an important part of sheep management. It is estimated that depredation, the loss of sheep life from predators, accounts for 15 percent of the total cost of sheep production in the United States, which is second only to feed and pasture costs. In 2004, the National Agricultural Statistics Service (NASS) estimated that 224,200 head of sheep and lambs were killed by predators, but the number could be as high as 275,000.

For centuries, shepherds closely guarded their flocks against predators. More recently, livestock guardians such as dogs, llamas, and donkeys have been used as well as protective barriers to ensure the safety of sheep.

SHEEP BEHAVIOR

Sheep are a prey animal, which means they are sought as a food source by other animals. Sheep do not prey on other animals. Their natural instinct is to flee in the face of danger rather than stay and fight because they have no means of protecting themselves.

Sheep employ an avoidance-and-rapid-flight response to any perceived threat. This allows them to use their natural herding instinct to congregate and face their adversary. No matter in which region of the country sheep are raised, they are most vulnerable to attack when they are alone and away from the rest of the flock. In this respect, breeds that have a strong flocking instinct seem less vulnerable to predators than those that might scatter.

One of their cautionary behavior traits is to seldom walk in a straight line. While their movement patterns in a field may seem erratic, their winding trails allow them

Sheep have a natural herding instinct and will congregate and face any perceived threat. They may not be able to thwart an impending attack but their movements can alert you to the presence of potential danger.

148

to keep watch behind them. Their excellent vision allows them to spot danger from 3,000 to 4,500 feet away.

Sheep have an excellent sense of smell, which they use to detect the scent of different predators. They also have excellent hearing and the movements of their small ears allow them to detect sounds emanating from sheltered areas.

Although they may not be able to thwart an impending attack, their safety-in-numbers mentality leading to a herding situation can alert you to the presence of potential danger to them. Routine observation of your sheep will help you recognize signs of abnormal behavior patterns.

COMMON PREDATORS

Some sheep producers encounter few or no problems with predators, while others may have to contend with the problem on a regular and consistent basis. A sheep predator can be any animal that attempts to kill them for food. This can include coyotes, bears, bobcats, foxes, wolves, wild or free-ranging dogs, and mountain lions, depending on the region in which you live.

In a strict sense, predators that do not harm livestock are not a concern and are part of the balance of nature to help control rodent or deer populations. Problems arise when their regular food supply is short and they realize that livestock are easier to successfully hunt than animals in the wild. The most prevalent predators are coyotes, which account for 50 to 60 percent of all sheep losses. This is followed by wild or free-ranging dogs at about 25 percent; bears, mountain lions, bobcats, eagles, foxes, wolves, and vultures account for most of the rest. Each predator species has its own particular traits.

Coyotes are easy to recognize with their long snouts, erect ears, and bushy tail. They are often mistaken for light-colored dogs because their weights can range from twenty to forty pounds. Coyotes are efficient predators and will attack sheep at their throat and kill by strangulation or by severing the jugular vein. They will occasionally pull the sheep down by attacking the side, hindquarters, and udder. Attacking the throat enables a silent kill as the victim cannot bleat or make noise with its throat shut. A coyote kill is typically identified by the removal of the rumen and intestines from inside the

Coyotes can be persistent and clever. They may work in groups to draw a protector away from the flock, which allows other coyotes to track down their evening's meal. *Shutterstock*

carcass and sliding the carcass around as they eat. A young coyote kill may look like a dog's in the indiscriminate way in how and where they attack.

Foxes feed primarily on rabbits, small rodents, poultry, and other small animals and birds. They do not necessarily attack sheep unless they are small lambs and will try to carry them off to consume them rather than eating them at the site.

Bears are not commonly seen as predators in many regions of the country. They can become a problem in certain areas where their population has increased and their hunting areas are on the fringes of sheep pastures. *Shutterstock*

Foxes aren't as severe of a threat as other predators. They generally attack only small lambs. If this occurs, they typically carry the lamb off to feed on it under cover. *Shutterstock*

Bears and mountain lions are not common predators in many areas of the country, but when they do attack sheep they may consume their kill almost entirely and leave behind only the skin and large bones. They may kill more than one animal in a single episode but may not feed upon them all.

Wolves have become an increasing concern for farmers located in areas bordering Canada because of the rise in their population during recent years. Wolves have been protected for many years to prevent their extinction and now pose a higher risk for small-animal farmers living on the edge of their territory. Wolves will typically stay within their cover areas and will hunt in packs to attack larger animals such as caribou, moose, elk, and cattle. They usually bring down their prey by cutting and damaging the muscles and ligaments in the back legs or by seizing the victim by the flanks. Wolves have an uncanny ability to spot the slightest limp or other weakness when they are selecting prey and will choose the sheep they sense is easily captured.

Cougars, bobcats, and lynx occasionally prey on sheep and goats if they are within their range, but wildcats commonly kill smaller animals such as poultry or rabbits. Bobcats will use their claws to capture a sheep and often kill by a bite through the top of the neck or head. Cougars, bobcats, and lynx kills are often dragged some distance from the point of attack and may partially cover their kill with leaves or twigs.

Bald eagles have been reported as occasionally attacking sheep. If they do, they typically will select young lambs and skin out the carcass and may bite off and swallow the soft ribs. Being a scavenger bird, eagles are more likely to feed on carcasses left behind from other kills.

Wild, free-ranging dogs or domestic dogs may be more of a problem killing sheep than many of the predators mentioned because of their former domestication. Wild dogs frequently attack in packs, and whether single or in a pack, they tend to run through a flock, maiming as many animals as they can catch. The characteristic bite marks are on the flanks, rear legs, backs, and rear ends of the animals. Even if dogs don't kill or maim any sheep during their attack, the sheep may be in shock from being chased and succumb to the residual effects such as exhaustion or injury. Feral dogs can be vicious in their attacks and should be treated with caution.

SHEEP PROTECTION

There are several precautions and controls that you can use to safeguard your flock. While no system can be considered completely foolproof, you can minimize your risk of loss by predators by using fencing controls or employing livestock guardians to discourage them.

Observation is the simplest way to assess any potential predator problems. If you notice animals scouting your flock and are not sure whether they are coyotes or dogs, you may want to look for footprints or scat. Dog footprints are round and all four claw marks are visible, while coyote tracks are distinctly oval and only the front two claw marks are visible.

Developing a fencing program where the fences are difficult to breach may be an effective deterrent to predators. Predators like coyotes are smart and persistent. They have the ability to find holes in fences and can fairly easily jump a four-foot woven-wire fence. Coyotes can also dig under fences to gain entrance to a field. A five-foot-high fence with two strands of barbed wire above will serve as an effective barrier providing they can't dig under it. One disadvantage of this approach is that it may be impossible or too expensive to build a coyote-proof fence around your entire farm. An alternative is to maintain a coyote-proof area around the farmstead

Livestock guardians are used to defend your animals against predators. Llamas, donkeys, and dogs are most often used for this purpose. The best guard animals stay with the livestock without harming them and aggressively repel any predators. A llama's innate curiosity and size are intimidating to most would-be predators or scavengers.

Llamas have been used successfully as guardians because of their natural aggression toward coyotes and dogs. Their response toward predators is quick and assertive.

where you can pen your sheep at night. Lights may also help deter coyotes from approaching at night.

Electric fencing may help deter predators and can be built along existing fences. Predators tend to explore with their noses close to the ground and will typically approach a fence at ground level or slightly above. Using an electrified fence may serve as a deterrent but it must be maintained to retain its effectiveness. A dead electric wire will not deter any predator. Fencing can be more effective when used with livestock guardians.

Although electric fences or woven-wire fences topped with barbed-wire strands will be an effective deterrent for predators such as coyotes or foxes, they will not provide much protection from larger predators such as bobcats, bears, or wolves, especially if your farm is located in a wide-open area. Penning your sheep may be the only consistent solution to this problem.

Some producers have used noise-making devices as a means of predator control. While they may work initially, they have the disadvantage of conditioning any regular predator patrolling the area to their sound to render them ineffective, and may even have the opposite effect of attracting them to the sheep. Once a predator has developed a taste for sheep or chickens after discovering that fences are easy to breach and penned sheep are easy to kill, it will be difficult to keep them away unless other means are used, such as protecting your flock with guard animals.

LIVESTOCK GUARDIANS

Livestock guardians are animals that can display aggressive behavior as a deterrent toward intruding predators and may include guard dogs, llamas, donkeys, and cattle. The choice depends on the size of your flock, your location, the size of your farm, predator threats, budget, and personal preference. Depending on the animal you choose, they will need to be cared for in a manner that keeps them healthy and alert to any potential threats. This may include considerations for feed, training, veterinarian care, and housing costs.

There are two types of sheep dogs—guardian dogs and herding dogs—and each has different behavior patterns and objectives. Guardian dogs are used to protect sheep from predators, while herding dogs are used to

If you select a llama as a guardian for your flock, you should look for one with independence, curiosity, and awareness of surroundings. It should be wary but not afraid of dogs. Because llamas are ruminants they can be fed the same diet as the sheep they guard.

Donkeys can be used as livestock guardians and are growing in popularity because of their intelligence and acute hearing and sight. They will drive off a predator by braying, baring their teeth, and chasing while attempting to bite and kick.

manage sheep. Livestock guardian dogs are often referred to as sheepdogs since they most often have guarded flocks of sheep. However, they are capable of guarding other species of livestock. Guardian dogs originated in Europe and Asia where they had been used to protect livestock from predators for centuries. The most widely known breeds in the United States are the Great Pyrenees, Anatolian Shepherds, Akbash, and Maremma.

Guardian dogs are generally large, weighing seventy-five to one hundred pounds, and protective, but still can be gentle. They make good companion dogs and are often protective toward children. With the right socialization and training they can be used for family pets and home protectors. Because of their size, they are generally intimidating and typically remain aloof toward strangers. They typically don't reach maturity until about three years of age.

The purpose of a guardian dog is to bond with the sheep, and not the shepherd or owner, so that their protective instinct is for the sheep and little else. To be effective they must be courageous in the face of a predator and they must accept the responsibility of their job. Training for a guardian dog begins as a puppy, when they should be placed with lambs to develop an imprint on the sheep. This imprinting process is thought to be largely through the sense of smell and occurs at three to sixteen weeks of age. They tend to be species specific as a guardian pup raised with sheep will not be effective with goats or cattle or the other way around. This imprinting appears to be critical and should be accompanied with appropriate training. A good guardian dog will never hurt the sheep it is protecting. They live with the flock they are guarding and bed down with the sheep.

As a guarding dog matures, it spends much of its time near the sheep and keeps watch for outside animals that may approach the flock. If it chases an intruder away, it will return to the flock. Guardian dogs that follow the flock will be on hand if a predator attacks. A successfully trained guardian dog will find that the flock will surround him or her during the period of a perceived threat and will stay close until that threat dissipates.

Herding dogs differ from guardian dogs in their purpose. Herding dogs are not typically used for guarding the flock, although they may serve that purpose if properly trained. Herding dogs, such as Border Collies, are used to move the flock from one position to another. Their training is different from guardian dogs and they often cannot work both requirements.

Llamas have been used successfully as guardians because of their natural aggression toward coyotes and dogs. A tall, alert, three-hundred-pound llama can be intimidating to a coyote. Because they are ruminants, llamas can eat the same diet as the sheep.

Donkeys have been used for centuries to guard sheep and other herding animals. Donkeys are extremely intelligent and have acute hearing and sight to detect any intruder. They are protective of the flock and because their size towers over most predators, they are effective deterrents. They also eat the same forages as sheep, so no special supplementary feed is needed to keep donkeys on your farm. Only a gelded jack or jenny should be used as a guardian since intact males can be aggressive toward other livestock.

An often-overlooked livestock guardian is cattle. Placing sheep and cattle together in the same pasture tends to reduce predator losses. However, these different species must be encouraged to bond together, otherwise they will tend to graze in separate areas of the pastures and the effectiveness of the cattle guardianship is lost. Young lambs can bond with cattle by penning them in close confinement to cattle. When they are placed on pasture together, the lambs will follow the cattle. When a perceived threat from a predator appears, the lambs will stay close to the cattle.

LOSS AND COMPENSATION

The laws covering predator elimination may vary between local governments. You can contact your county agricultural extension agent for advice on how to handle situations that may arise in your area. You typically have the right to protect your animals if they are being chased by predators, including neighboring dogs. However, you will need to follow local, state, and federal laws governing predators.

If you happen to lose some of your sheep to predators, you may be eligible for compensation through government programs that address this issue. In the event of loss from predators, be careful in how you dispose of the remains. By not properly disposing of the carcass, you may be inviting the predator back. One of the easiest methods of disposal is burial, but this should be deep enough so that the carcass cannot be dug up or uncovered by carrion-eaters, some of which are capable of digging down several feet.

Guardian dogs are different from herding dogs, such as Border Collies, and pet breeds. All good guard dogs are characteristically courageous, fearless, large, strong, loyal, intelligent, self-assured, and capable of making independent decisions and taking action based on those assessments. However, proper training is an essential part of having a useful guard dog.

HELP AND ADVICE

Certain predator control practices may be distasteful, such as shooting, trapping, and snaring, but it may be the only way to remove individual predators, particularly if they start killing your sheep. Although new and better predator management techniques are being investigated, you should develop a plan that suits your needs and comfort level before it is needed. Develop a predator management plan with your local Department of Natural Resources office or county agricultural extension agent.

Grazing cattle with sheep may be a sufficient deterrent to predators because of the size of cattle and their natural protective traits. One challenge is to get both groups to bond with each other so that the sheep will seek protection from the cattle.

MARKETING
MEAT, WOOL, AND CHEESE

Your end options for raising sheep can include selling them for meat, selling their wool, and producing cheese, in the case of milking sheep. Your goals will likely decide which direction you take. To be successful in any of these markets, you should spend time researching your marketing options, plan your marketing strategies, and know the prevailing market prices.

MEAT PRODUCTION

The meat from sheep up to one year of age is referred to as lamb and is usually taken from a carcass weight of fifteen to sixty-five pounds. This includes both halves of the body, as well as the bone and fat.

A number of options are available if you want to raise sheep for meat sales. These options can involve several different production systems, including selling lambs at sheep markets, selling lambs as feeders to other growers, or raising lambs yourself to market weight. Selling lambs directly to consumers involves raising lambs to a specified weight before having them processed to sell.

Meat from sheep, particularly lambs under one year of age, is finding an increasing market in restaurants, grocery stores, ethnic food outlets, and the everyday consumer. It is high in protein and conjugated linoleic acid (CLA), a unique and potent antioxidant naturally produced through pasture grazing. *Steve Pinnow*

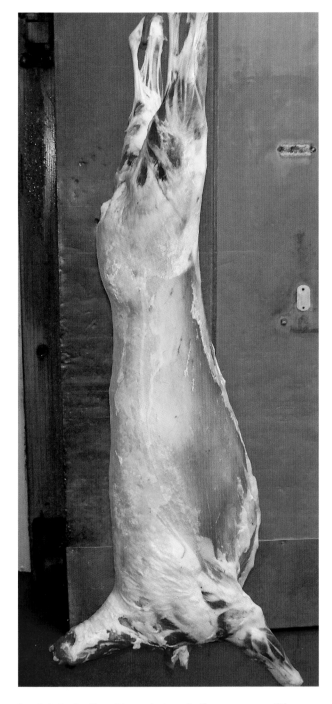

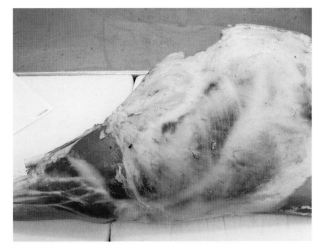

Although a lamb has four legs, only the two hind legs produce the cut referred to as leg of lamb. It is a large, lean, and tender lamb cut and can be used whole or subdivided into smaller lamb cuts, which can be prepared in many different ways. *Steve Pinnow*

Lamb is typically sold as whole or half carcass quantities or as single cuts. Marketing meat products from your farm to consumers is one option to add value to your program. There are rules and regulations regarding the processing, storage, and handling of fresh meat that is to be sold to the public. You can learn about these requirements by contacting your county agricultural extension office. *Steve Pinnow*

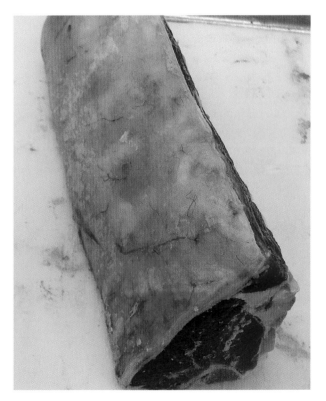

The lamb loin primal cut is the section along the lamb's back from the thirteenth rib to the hip. The lamb loin contains the most expensive, highly prized, and tender meat. *Steve Pinnow*

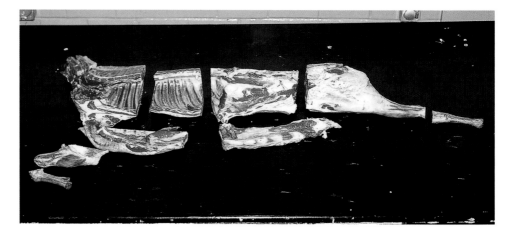

There are five primal cuts on a lamb carcass. They include the shoulder, rib, loin, breast, and leg. Each primal cut can be divided into a variety of subprimal and market ready or ready to cook cuts. *Department of Animal Sciences, University of Wisconsin-Madison*

The carcass of a sheep is often divided into three kinds of meat: forequarter, loin, and hindquarter. The forequarter includes the neck, shoulder, front legs, and the ribs up to the shoulder blade. The hindquarters include the rear legs and hip. The loin includes the ribs between the forequarter and hindquarters. Lamb is one of the richest sources of conjugated linoleic acid (CLA), which is a unique and potent antioxidant produced naturally from linoleic acid by bacteria in the stomachs of herbivores such as sheep and cows.

WOOL PRODUCTION

The wool from one sheep is called a fleece, and according to USDA statistics, the average yearly fleece weight in the United States is 7.4 pounds per sheep. In 2006, the U.S. total wool production was approximately 36 million pounds. Wool is a renewable resource because sheep produce this unique fiber through the conversion of feedstuffs that might otherwise be wasted.

Wool has many uses, and the most prevalent is in making clothing. Wool is resilient, lightweight, and has excellent insulating properties. Sheep producers traditionally harvest wool during the spring months, before the summer heat arrives. More than half of the wool production in the United States is shorn and sold during April, May, and June. Marketing wool can be done through several different methods, including outright sale to brokers or by transforming it into a product that can be used by hand spinners, knitters, and weavers. In the past, textile mills in the United States consumed nearly all of the domestic wool production with process-ing plants predominantly located in the eastern states, such as Vermont. However, in the past decades many of the mills have either closed or moved their production facilities to other countries. This shift has made export markets an important segment for U.S. wool producers.

WOOL CHARACTERISTICS

The characteristics of wool differ between breeds and breed types and are controlled by the genetics specific to that breed. Long wool produces the heaviest fleece due to the size of the fibers. Hand spinners prefer long wool because it is easier to spin. The lightest-weight and

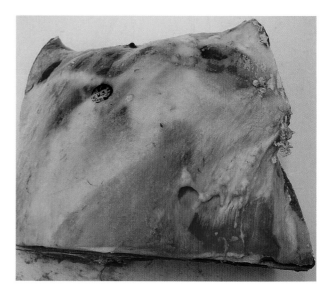

The retail cut taken from the shoulder can include the arm and blade chops. It can be cubed for kabobs or cut for rolled shoulder or square shoulder roasts. *Steve Pinnow*

least-valuable fleeces are typically produced by medium-wool sheep, which are raised more for their meat than fiber. Medium-grade wool is often used in making blankets and sweaters. Fine wool is the most valuable because the small fiber diameters have a greater versatility for use. Coarse fibers are known as carpet wool and are generally used to make carpets, rugs, or tapestries. Wool typically is packed into large square bales or large wool sacks that can be easily transported.

VALUE OF WOOL

The end use of the wool will largely determine its value. Raw wool is usually purchased on a grade basis that categorizes the fiber length and diameter. The grade of wool is discounted if it is dirty or contains a lot of organic matter or contaminants such as bedding materials, feed particles, manure, or other things that can adhere to the fibers. White wool is more valuable because it can be dyed any color, while colored wool cannot. Fleeces that contain mixtures of colors, such as white wool from the body and black wool from the face, are less desirable because the black fibers cannot be dyed.

The price of wool can fluctuate. During some cycles the price received for wool may barely cover the cost of shearing, while at other times it may be sold for a profit.

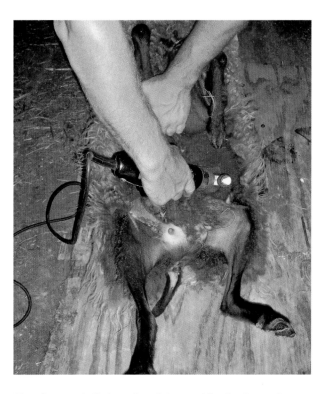

Shearing wool off sheep is painless and begins by setting the animal upright. This immobilizes it and allows the shearer to easily handle the sheep while using an electric clippers. The longest possible cuts are made starting at the neck and moves down the belly.

Wool is one of the most versatile products derived from sheep. It has many uses and its color is determined by the breed. Wool color can be changed by using different dyes during the processing from raw wool to yarn ready for weaving, knitting, or sewing.

As the sheep is gently rotated, the shearer cuts along the side, legs, and back to allow the removed fleece to fall to the floor. Most sheep do not move during the shearing process but you should be aware that their feet can cause injury if your head gets in the way if they kick.

Shearing leaves a fleece that can be processed. It is typically rolled up while other sheep are sheared. Some growers prefer summer shearing over a spring cut because of showing events. Summer-sheared wool tends to be cleaner and contains less chaff, dirt, or vegetable matter.

The finished sheep suffers no ill effects from being sheared and will readily grow another coat of wool. It is a renewable yearly product to be harvested. There are shearing seminars where you can learn the art of shearing sheep. *Department of Animal Sciences, University of Wisconsin-Madison*

Wool can be sold privately to hand spinners and knitters through brokers or wool pools. Large producers of wool typically sell theirs to warehouses or directly to wool mills. It can be sold through wool pools, which are a means of bringing together smaller volumes of wool to improve the marketability of the wool through larger lot sizes. There are more than one hundred wool pools located throughout the country and they are a common venue for small producers of wool. Contact your county agricultural extension office for more information about local wool pools.

FROM WOOL TO FABRIC

There are five steps involved in creating fabric from wool: shearing, scouring, carding, spinning, and knitting/weaving. Shearing removes the wool from the body of the sheep by cutting or shaving and can be likened to giving them a haircut. Most breeds grow wool continuously so it is important to remove it at least once a year. Most sheep are sheared in the spring, but some are sheared prior to lambing to make it easier for newborn lambs to nurse. Sheared sheep take up less room at feeders and in the barn. Shearing is typically done using an electric clipper.

The process of removing undesirable parts of the fleece is known as skirting and is the first step in processing the wool. Wool will typically contain stains, coarse wool fibers, and vegetable matter such as weeds, burdock seeds, and other things that lower the quality.

A simple mechanism for breaking apart clotted raw wool is called a triple-picker. It has sharp steel nails embedded in a solid wood frame and is an effective method of loosening sheared wool.

Scouring is washing the fleece in tubs of water to remove dirt, grease, and other foreign materials using alkaline solutions such as soda, borax, or laundry powder. Carding prepares the raw or dry fibers for spinning. It aligns and straightens the different fibers to create a common mix and eliminates the tangles. Carding can be done either by hand or by machine.

Spinning is turning the wool into yarn. Coarse wool is spun into woolen yarn, while fine wool is spun into worsted yarn. Woolen yarn is typically used in carpets or thick sweaters, and worsted yarn is used to make lightweight

Once pieces of foreign matter have been broken up or removed from the wool, it can be placed in a washing machine that has no agitator. This will wash the wool and remove the dirt and lanolin without subjecting the wool to felting, which occurs when wet wool is spun and matted together rather than allowed to dry.

fabrics for suits and dresses. Knitting and weaving, either by hand or machine, turn the yarn into fabric.

WOOL PROGRAMS

The national contest for Make It Yourself With Wool is currently sponsored by the American Sheep Industry and is an outgrowth of a 1947 cooperative effort between members of the Women's Auxiliary groups of the sheep growers organization in the West and 4-H clubs. Today, anyone with an interest in creating clothing from wool is eligible to compete. They offer four age classes from pre-teens (age twelve and under) to adults (age twenty-five and older) where the contestants must select, construct, and model their own garments.

The program promotes the beauty and versatility of wool fabrics and yarns and encourages personal creativity in sewing, knitting, and crocheting woolen garments. In the most recent state contests, 926 junior and senior contestants used 2,800 yards of wool fabric and 952 skeins of yarn to create their garments. The most recent national contest had sixty-two junior and senior finalists representing thirty-one states.

Washing does not hurt the fleece and you can use a mild liquid detergent to break down the lanolin on the wool. It is important that the wool is completely clean before it is dried.

SHEEP MILK AND CHEESE

Sheep milk accounts for about 1.3 percent of the world's total milk production and most of it is made into cheese, including the well-known feta, ricotta, and Roquefort. Sheep milk is highly nutritious and richer in vitamins A, B, and E; calcium; phosphorus; potassium; and magnesium than cow milk. It is also more easily digested than cow milk because it contains smaller fat globules. Sheep milk can be frozen and stored without affecting the cheesemaking qualities. This allows the milk to be stored until a sufficient amount is available for transport, sale, or cheesemaking. Because of the higher solids content in sheep milk, more pounds of cheese can be made than from an equal volume of cow or goat milk. Sheep milk will yield 18 to 25 percent cheese, while cow and goat milk typically yield about 10 percent. Although you will get significantly less milk from sheep than from goats or cows, sheep milk can sell for four times the price of cow milk and the price will also be reflected in the price of the cheese. While whole or fluid sheep milk is used in cheesemaking, it typically is not consumed as a drink. Sheep milk can also be made into ice cream and yogurt for value-added products.

Artisan sheep cheesemaking is becoming more popular with consumers. Selling sheep milk may be one way to increase the productivity of your flock. However, there will be a higher initial investment because the milk must be refrigerated until it can be transported to a cheese factory. If you decide to make cheese yourself, you will need to invest in equipment to make the cheese.

Some sheep-milk producers become licensed cheesemakers and create distinctive cheese brands. Sheep milk

Washed wool must be laid out to dry before carding. A rack, along with warm air, can be used to dry the wool.

A hand-turned carding machine aligns the fibers as they are fed into the drum. This straightens the fibers and the carded wool can be peeled off the drum, laid in flat sections, and combined with other wool. Different colored wool can be mixed in during this process to create unique color combinations.

can be mixed with cow or goat milk to create cheeses that are value-added products. If you are going to pursue a cheesemaking program, you will need equipment in which to process it and a storage area once it has been made. There are licensing requirements for producing food; you should contact your state Department of Agriculture to learn more about their rules and regulations pertaining to the production of a food product if you plan to market it.

Most sheep outside of the United States are milked by hand because they are often raised on land where cows cannot survive. Modern U.S. sheep dairies have machines similar to those available for dairy cows, including milking machines, parlors, pipelines, and bulk tanks.

DAIRY SHEEP BREEDS

Ewes of any breed can be milked, but there are several dairy sheep breeds that produce more milk per year than other breeds. These include the East Friesian, which is the most common and productive breed of dairy sheep in the world. An East Friesian can produce an average of 1,000 pounds of milk per lactation (length between lambings), which lasts 220 to 240 days. This makes an average of about 4 pounds of milk per day for seven to eight months.

Dairy ewes can be managed in different ways depending on the farm goals. Some ewes are not milked until their lambs are weaned at thirty to sixty days and are then milked twice every day. A different management

Clean, dry wool can be safely stored in plastic containers for later use. Different color dyes can be used for creative and artistic clothing creations. If wool is stored it is very important that it has been sufficiently cleaned, dried, and carded.

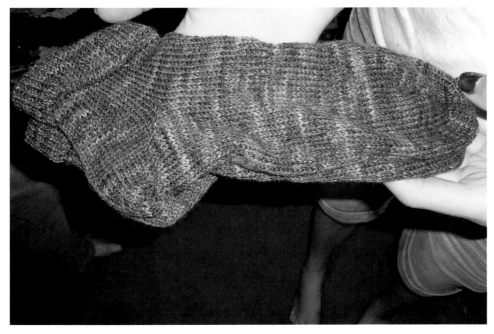

Finished products, such as colored socks and sleeveless vests, provide warm clothing and make terrific gifts. An important industry has developed around hand spinners across the country.

There are specialized dairy sheep breeds much like there are specialized breeds of cattle and goats for dairy production. The East Friesian and Lacaune (shown here) are the most prevalent in the United States and can produce from 400 to 1,100 pounds of milk per lactation.

system allows lambs to nurse for half the day, after which time they are separated from their mothers for the evening and the ewes are milked the following morning. A third option is to wean lambs at one month of age and milk the ewes twice daily after that period.

The greatest volume of milk for the entire lactation is when the lambs are removed from their mothers within twenty-four hours after birth and raised on an artificial milk replacer. If you are considering this option, you will need to weigh the benefits of the increased milk supply from the ewes against the added cost of purchasing a suitable milk replacement product and the time spent bottle-feeding the new lambs.

SHEEP CHEESE PRODUCTION

Sheep-milk cheeses are appreciated for their sharp, bold flavors. If you want to make cheese from sheep milk you will need to determine the number of sheep required to make it worth the time and effort. This involves developing a business plan to determine all the costs involved and expected returns.

Producing a product such as sheep cheese is not necessarily the difficult part. With training, equipment, and experience, many people can develop an expertise at making good cheese. Perhaps the biggest challenge is to develop a market for your product prior to creating it. Advance research and study of your target market will help you determine your direction. While there is always a certain amount of risk inherent in starting and developing a new food product, laying the groundwork for sales, whether private or for market outlets, will give you a better foundation for success. There are many university and extension programs available to help you set up a cheesemaking business; develop a business plan; and provide direction in licensing, regulations, inspections, and marketing strategies.

Right: Sheep milk cheese is gaining in popularity with consumers. It is nutritious and provides a distinct alternative to cow milk cheese. You can milk sheep with equipment adapted to that purpose. There are regulations relating to storing and processing sheep milk and you should learn more about them if you wish to pursue this market.

The Dorset and Polypay (shown here) are two U.S. breeds that are best adapted to dairy production. These breeds produce between 100 to 200 pounds of milk per lactation. Crossbreeding domestic breeds with specialized dairy breeds can boost production to an average of 250 to 650 pounds of milk per lactation.

ENVIRONMENTAL STEWARDSHIP AND NUTRIENT MANAGEMENT

Sheep can have a positive impact on the environment by grazing areas not accessible to mechanical harvesting where noxious weeds can grow. The sheep spread their manure while grazing, which fertilizes the grasses. *Cynthia Allen*

Raising sheep can have a positive effect on the environment. They can be used to help control several effects upon pastures and grass management, including nutrient management, weed control, riparian and wildlife enhancement, silviculture, and waste management.

ENVIRONMENTAL STEWARDSHIP

Environmental stewardship involves the influence your farming activity has upon your land as well as your responsibility to maintain viable and healthy surroundings for future generations. This stewardship involves matters beyond the primary concerns of proper application of livestock manure, such as controlling odor and maintaining aesthetics to make your farm inoffensive to others living nearby. It also includes decreasing soil erosion, watershed cleanliness, and invasive weed control.

This stewardship, besides embracing the land, air, and water resources, also considers the relationship of one neighbor to another. Sound science will assist producers in managing their farm in an environmentally responsible manner, while sound ethics will ensure that producers conduct a humane animal agriculture. Good stewardship practices can make farms more profitable and reduce petroleum product inputs such as purchased fertilizers that may not be needed with the proper application of farm manure.

SHEEP ASSISTANCE

Sheep can be used to assist any program devoted to environmental stewardship because of their low impact upon their surroundings and their ability to be used in management practices such as enhancing wildlife habitats.

Grazing sheep is a relatively low-cost, low-impact endeavor, although sheep have different dietary needs and grazing habits than other livestock. Because of this difference, sheep can be useful in developing a multispecies grazing program. This program can result in plant communities that are more resistant to grazing impacts by only one species and other factors affecting the ecosystem stability, such as drought.

Multispecies grazing uses resources more uniformly and sheep have the ability to neutralize plant toxin that may be lethal to cattle. Sheep can graze such plants as larkspur, leafy spurge, and ragwort without any adverse effects. Grazing sheep over infested areas can reduce the risk of cattle being poisoned and allow other plant species to grow. Using sheep to control these plants can minimize or eliminate the use of chemicals. Some studies have shown increased weight gain in both species because of the increased carrying capacity of the land and from better individual animal performance.

CONSUMERS OF WASTE

Sheep can be used to convert waste products into wool, meat, manure, and milk for cheese by using their digestive systems to process crop residues such as corn stalks or turnip leaves, which are unfit for human consumption. In some instances, urban parks have found sheep to be an economical lawn-mowing service, which reduces the need for petroleum-powered equipment.

Wool has been found to be an excellent absorbent, acting like a sponge, for petroleum product spills such as diesel and crude oil. One private study has found that wool can absorb twelve to forty-four times its weight in oil, depending on the pad thickness and type of oil. This may be useful in oil spills that occur in natural habitats where other machines might not be able to enter. Wool products for oil-spill cleanups have been used in Europe and Australia since 1990 and are now breaking into the U.S. market.

WATERSHED MANAGEMENT

In watersheds or low areas on your farm where shrubs dominate, sheep can be used to help manage vegetation and decrease soil erosion. Their ability to clear brush through grazing can promote the growth of perennial grasses that enhance watersheds. Their pointed hooves can puncture the soil structure and increase the ability of seeds to germinate and grow. The United States Forest Service uses sheep grazing effectively for riparian restoration and watershed recovery.

Your farmland may contain waterways, streams, ditches, or ponds. Seeding the banks may enhance soil retention but also provide areas for plant growth that can't be readily harvested. Sheep can be used to graze these overgrown areas to prevent unwanted plant species from establishing themselves. Sheep do not willingly enter waterways, so they can be easily pastured along stream banks without damaging the soil structure. You may have areas on your farm that can be restored by using sheep, although these areas may be projects that require several grazing seasons.

As ruminants, sheep can process crop residues and other plant materials not fit for human consumption and convert them into useable products such as meat, milk, and wool.
Cynthia Allen

Manure from healthy sheep has the appearance of rabbit droppings or little round balls. As these manure clumps decompose, they can provide nutrients for the plants and homes for insects that feed on fibers and organic materials. These insects in turn provide food for many grassland birds.

NUTRIENT MANAGEMENT

Effective waste management is one of the most important aspects of any animal-based agriculture production program. Good stewardship dictates that manure be handled in a responsible manner that enhances rather than damages the environment. Improper utilization of manure or poor management of animal wastes can be detrimental to water quality and affect a wide range of species. A nutrient management program ensures that manure components are used in a way that provides nutrients for your soil through balanced applications across your fields while minimizing the detrimental effects on the surrounding environment, such as waterways, streams, or other water sources, including wells.

Manure is an inevitable by-product of sheep production. How it is used will largely determine if it is a valuable asset or a costly liability to your farm. Depending on the number of sheep you plan to raise, you can develop a nutrient management plan before any animals arrive on your farm. Planning ahead will make the utilization of manure more effective in plant nutrition and lower your crop fertilizer costs.

Water contamination from manure typically occurs as a result of runoff during times of heavy rain or when manure is spread on frozen ground where it has little or no chance of being absorbed into the ground prior to rain or snow melt. Farm waste runoff into streams, creeks, ponds, lakes, and other water bodies has become one of the most contentious issues between rural landowners and urban populations that view these waterways as recreational areas. Well-planned use of manure can avert problems before they occur and can be a valuable asset instead of a liability to your farm.

MANURE PRODUCTION

The normal digestive processes of a sheep's gastrointestinal system produce fecal material and urine that is expelled as manure. It quickly decomposes under warm, moist soil conditions and releases nitrogen, phosphorus, potassium, and other nutrients into the soil.

While manure was recognized very early in human civilization as having beneficial effects on crops and plants, it also has had various uses not related to soil fertility, including fuel and shelter construction. With the introduction of commercial, petroleum-based fertilizers, manure became a problem to be disposed of rather than an asset to be utilized. Today, the emphasis is on keeping manure runoff from reaching water sources and using it to help improve soil fertility, structure, and composition. Field plants absorb nutrients from the soil to grow. Replenishing these nutrients with manure has been a solution to lowering purchased fertilizers.

Solid manure, when mixed with bedding such as straw or hay, can add fiber and organic matter back to the soil. This combination, when plowed or stirred back into the field, can loosen soil particles and allows the soil to become more porous and absorb more moisture, and creates more water-holding capacity. Loosening the soil particles relieves compaction of the soil structure, which allows plants to develop a better root system for better growth.

CALCULATING MANURE QUANTITIES

The amount of manure produced on your farm can be calculated fairly accurately. A one-hundred-pound

In pasture grazing programs, the manure is spread by sheep moving around the paddocks. With good pasture rotation your fields can benefit from manure application with little mechanical assistance.

market lamb will typically produce about four pounds of manure and urine each day. Multiply that by the number of mature sheep you have and you will have the approximate daily volume of manure produced.

If you have only a few sheep, the amount of manure produced each day will not be great. However, if you have one hundred ewes, the total will add up over a period of several months if they are not on pasture.

Although the nutrients in manure can be considered beneficial if used properly, they can also become pollutants if they enter streams or groundwater systems and have serious effects on wells and the quality of drinking water. Understanding the proper handling of manure will help you avoid potential causes of water pollution.

MANURE HANDLING

There are three types of manure classes: solid, semisolid, and liquid. Although each type needs to be handled differently, sheep typically produce solid manure that can be easily handled. Solid manure consists of a combination of fecal matter and urine that has been mixed with dry bedding materials, such as straw, hay, wood shavings, corn

fodder, and any other materials. Solid manure is typically handled with a manure spreader that applies the material on croplands or pastures to be used as fertilizer.

The most economical way of spreading the majority of the manure produced by sheep is to let them distribute it themselves in the pastures while they graze. If there is enough pasture area to feed the animals and still sustain the vegetation, the solid manure produced by each sheep is more or less spread uniformly around the pasture as they move about and requires no extra handling on your part. Any possible runoff effects from pastures can be minimized by rotating pasture areas. The manure sheep create will slowly decompose in the pasture and provide a habitat for a wide range of useful insects that help break down the manure into nutrients used by the soil.

MANURE STORAGE OPTIONS

Farms with large animals will have different manure storage capacities than one with sheep. With fewer, smaller animals your storage options can be affordable because the startup costs are usually minimal.

Composting is one way to handle manure in an environmentally sound manner and can be a potential

Composting and establishing and maintaining grass strips along streams are environmentally sound conservation practices that also provide wildlife habitats. Compost piles can be spread when weather and field conditions are better for nutrient absorption by the soil.

There are many advantages to composting including reducing the volume of manure and breaking down the volatile organic compounds into more stable forms. Composting also reduces manure runoff in fields and leeching into ground water. Insects that find a home in compost can become food for birds of all kinds.

revenue source for your farm. Composting is the active microbial treatment of solid manure by using oxygen as the main catalyst for this process. The organic matter is allowed to decay in a pile or windrow. Decaying organic matter creates heat and a compost pile of manure, depending on its density, can reach temperatures of more than 160 degrees Fahrenheit at its core. Oxygen is required for composting, so the pile needs to be turned or stirred occasionally for the material at the edges of the pile or windrow to become part of the heating process. A tractor with a loader or other equipment designed for stirring piles or windrows can accomplish this task.

Composting has the advantage of reducing the volume of manure and transforming it into a more stable nutrient form. These nutrients, when spread on the fields, are slowly released into the soil for crop nourishment. Because of this, the nutrients in composted manure are less likely to be transported off the site through runoff and leaching into the groundwater.

A second advantage of composting is that manure can be stored until weather and field conditions are better for hauling and the absorption rate of the nutrients by the soil is greater. Composted manure can also be sold as an off-farm fertilizer, soil additive, or mulch. Garden stores offer these products in plastic bags for gardeners, vegetable growers, and others. Because the manure has been broken down into less volatile nutrients, it provides a more benign product for the general public's use. The development of a market for composted manure may have more to do with distribution than with actual production. Information regarding this market and how to develop it is generally available from your county agricultural extension office.

MANURE MANAGEMENT PRACTICES

Farms expanding beyond a certain threshold for animal numbers or those involved with government farm programs are required to develop a nutrient management plan. Nutrient management plans are developed with professional experience and approval. These plans develop a program for each farm where the nutrients produced from manure are accounted for in the total field application of fertilizers, whether purchased or not.

Accounting for nutrient application on all fields ensures that excessive amounts are not used, thereby limiting the possibility of leaching or runoff into the groundwater supply. Areas of high vulnerability for runoff are identified and spreading manure in those areas becomes restricted. By not exceeding crop requirements, the soil is not saturated with nutrients it cannot absorb.

Soil tests are one part of a good manure management plan and contribute to the stewardship ethic by identifying those areas needing correction, as well as areas not needing any application of manure. These tests are the starting point for determining the nutrient content of your fields; the results explain the requirements of each field and will serve as a guide to application rates.

SOIL CONSERVATION PLAN

If you plan to participate in a federal farm program, a soil conservation plan is required. The conservation plan is a part of nutrient management programs because it identifies crop rotations, the slopes of all fields, and the conservation measures you will need to follow to stay within the tolerable limits of soil erosion.

This plan also identifies which fields may have restrictions for spreading manure because of close proximity to waterways, especially in winter. One component of this plan includes identifying the best time of year to spread manure. This will depend upon the manure-handling system on your farm. A farm with manure storage has a plan different from one that requires daily manure hauling.

Properly designed buffer strips along stream banks adjacent to fields with a potential for runoff can help prevent manure materials from entering the stream. These grassy strips help stop, filter, and hold back the sediments of runoff but still can be grazed by your sheep to maintain their usefulness.

Many problems with farm manure can be avoided with a plan that involves the best utilization of it regardless of the size of your flock. Nutrient and manure management plans and a conservation plan for your farm can be developed with help from your county agricultural extension agent or the Natural Resources Conservation Service.

HANDLING THE DEAD BODY

Death loss on a livestock or sheep farm is inevitable and producers must be prepared to properly dispose of the dead stock, whether the death came naturally or through euthanasia. There comes a time on every sheep farm when an animal suffers an injury, illness, or some other debilitating condition that requires euthanasia to provide a swift and humane death. This should be done in a manner that will minimize the stress and anxiety experienced by the animal prior to unconsciousness.

Correct euthanasia procedures produce rapid unconsciousness followed by cardiac arrest and total loss of brain function. There are several methods to accomplish this, including chemical, which is always administered by a licensed veterinarian using a barbiturate product. Sheep euthanized in this manner should not be used for human consumption or fed to other animals because of the residue left in the body. After all vital signs of a euthanized sheep disappear, the body must be disposed of.

A physical method that does not require human contact with the animal is shooting it with a gun. Strict firearm safety must be observed, as well as local laws and ordinances relating to the discharge of a firearm.

The Environment Quality Incentives Program (EQIP) provides incentive payments and shares for farmers to implement various conservation practices including fencing and filter strips. Limited resource producers and beginning farmers may be eligible for cost shares. Contact your county agricultural extension office for more information.

Disposing of a fallen animal is a biosecurity and environmental issue that is becoming more important and more difficult to handle. There are several acceptable ways of disposing of a dead animal and the one that fits your farm is an individual decision. Most states have regulations relating to the disposal of dead animals. Burials are allowed in most states but must conform to certain depths of burial, time limits, distances from wells, adjoining properties, waterways or streams, lakes, and residences, to name a few. There are about three hundred licensed rendering facilities operating in North America, but because there are geographical voids, it may not be possible to use one of them. Information on guidance and assistance, as well as ordinances in your area or state, can be obtained from your state department of natural resources or your county agricultural extension office.

BIOSECURITY AND COUNTRY LIVING

Biosecurity should not be confused with bioterrorism or agro-terrorism because they are not the same thing, although biosecurity is part of each concern. No matter what size flock you have, it takes only one sheep or one pair of dirty footwear to introduce a new disease into your flock. It also takes only one farm to initiate a disease epidemic.

As a farm owner, the responsibility for the biosecurity on your farm rests with you, along with the plans and steps taken to prevent the introduction of any infectious disease onto your farm and to limit the spread of any disease already present within your flock. To be successful, your plans must address how infected animals will be isolated from other animals in your flock and how cleaning and disinfecting procedures will be used.

AVOID INTRODUCING DISEASES

The greatest risk of introducing an infectious disease onto your farm is by bringing in new sheep that have been exposed to a disease or have it themselves. This is one reason that visual observation of any new animals is important as the first line of defense in the initial screening of the sheep. While they may appear healthy, they could be carrying a disease that could devastate

In the context of livestock production, biosecurity refers to those measures taken to prevent the introduction and spread of diseases on your farm. These measures exist on three levels: national, state, and individual flocks. Your flock is the area where you will likely have the greatest influence on the introduction of disease agents.

Biosecurity is an important measure you can take to reduce or minimize the introduction and spread of diseases on your farm. It involves maintaining a healthy flock and purchasing only healthy animals from breeders who share your common concerns. Biosecurity measures begin with you and your family being aware of contact you've had with animals or other farms every time you return home.

Isolate any newly purchased sheep for a minimum of four weeks and up to six weeks before allowing them into your pastures or to co-mingle with other members of your flock. This isolation will allow you to observe or detect any disease problems before exposing them to the rest of your sheep.

your flock. Avoid buying animals with abscesses, lameness, blisters on the mouth, abnormal breathing patterns, discharges from mouth or eyes, and anything else that doesn't appear normal.

There are several things you can do to diminish your flock's health risks, including knowing the health status of the flock you purchase animals from, isolating new purchases, limiting access to your farm, using a preventative health management program, and maintaining a closed flock.

If you are purchasing sheep, you should try to buy them from reputable sources where you can view their flocks and farming program. If possible, buy sheep from a closed flock. Purchasing animals at a sale barn may be an easy way to acquire the number of sheep you need, but it may bring in infected animals. The animals in a sale barn are exposed to other animals, and even though they may appear healthy, they will be exposed to whatever contagious diseases are present. Sale barns tend to attract cull animals that are not the best quality. However, this may be a venue in which you market sheep you no longer want as members of your flock.

It is important to prevent the introduction of scrapie to your flock, and you should try to purchase animals from USDA-certified scrapie-free flocks or flocks enrolled in the program. Purchasing sheep with scrapie-resistant genotypes, such as RR or QR, will also help prevent scrapie from entering your flock.

Whether you purchase animals in a sale barn or from private or public venues, be sure to isolate them from the rest of the animals on your farm for at least thirty to sixty days, and have your local veterinarian examine them for physical problems.

ISOLATION

Whether you purchase sheep through a public or private venue, you should isolate the new sheep from your flock to assess their health status prior to mixing them with the rest of your animals. The isolation time can vary from a minimum of thirty days up to sixty days, which will allow the incubation of most diseases to show symptoms in any exposed animal. During this time, it is important to observe the behavior and all vital signs of the sheep as they may indicate early clinical signs of the presence of a disease or infection.

An isolation area at least one hundred feet from the rest of the flock is recommended. Your new sheep should be confined, preferably in a separate barn or at least in a corner of your barn where they cannot have nose-to-nose contact with the rest of your flock.

Isolation pens allow you to feed newly purchased sheep separately to observe eating and drinking habits as they acclimate themselves to their new surroundings.

Your isolation areas should be away from the other sheep, and they should not be allowed to share the same airspace or feeding troughs with the rest of the flock. Generally speaking, the farther away your isolation area is from the others, the better. It is best to maintain a minimum of one hundred feet between the isolated animals and your flock.

During isolation, check the new sheep for external problems such as footrot, which is a disease that corrodes the hoof tissue and can eventually render the sheep lame, or cause other feet problems. It is important to keep footrot bacteria off your farm if none exists because the bacteria will live in the soil and can't be eradicated once introduced. The only way footrot can be transmitted to a farm that doesn't have the disease is through infected animals. Another external body concern is caseous lymphadentis, an infectious, contagious disease that affects the lymph system and is usually introduced through contaminated shearing equipment. This disease creates boils or abscesses when the skin is cut or ruptured. To prevent infections, be certain your shearers disinfect their equipment thoroughly prior to working on your flock. You can also check for ticks, warts, and

other external parasites during isolation and provide a way to eliminate the parasites before the sheep expose the rest of your flock.

When you bring new animals onto your farm and implement isolation procedures, be certain that this isolation also pertains to your feeding methods. It will not be an effective isolation program if you track your footwear through the areas of the new animals and then into your present flock without proper sanitation procedures. Your local veterinarian can assist you with developing an isolation area and correct procedures prior to bringing any animals onto your farm.

LIMITING FARMING ACCESS

Agritourism has become an attractive program for many small-scale producers who try to provide educational experiences for urban residents while creating another revenue stream for their farm. Balancing that with your need for biosecurity can pose challenges but does not need to be an obstacle to your plans. Having common-sense procedures in place can minimize your risk from outside sources. These procedures might include providing clear plastic, disposable boots for visitors to wear while walking around your farm or viewing areas. Disposable boots may be useful no matter who your visitors are. Unless the people coming onto your farm have been exposed to other sheep farms with their footwear, there should generally be little trouble with exposure to other sheep diseases from urban visitors.

Besides footwear, vehicles can be a source of contamination to your farm. While this has not been a major problem in the United States, it can be in the case of a local or regional outbreak of an infectious disease.

PREVENTATIVE MEDICINE

Preventative medicine is a biosecurity measure that is less expensive than curative medicine. Preventing health problems from entering your flock allows your animals to grow without having to contend with infections or other ailments. A vaccination program provides insurance against many common sheep diseases and is generally recommended for sheep and lambs. The use of vaccines will depend upon your or your veterinarian's perception of disease risk for your flock. Vaccines are available for many sheep diseases including vibro

Although showing sheep is useful and educational, it does pose a certain degree of risk by exposing your animals to other producer's sheep. Unless their animals are from the same flock, exhibitors should not share equipment, watering, or feeding pails with any other exhibitors.

Some diseases can be spread by contaminated footwear and vehicles. You can limit the introduction and spread of diseases by limiting access to your farm and flock. Requiring visitors to use plastic, disposable boots, especially if they have been on another sheep farm, is a good biosecurity practice.

and Chlamydia abortion, epididymitis, soremouth, and rabies, to name a few. Some of these vaccines should not be used until the presence of the disease appears and your local veterinarian can advise you on those that should be administered.

Sheep raised in warm, humid climates are susceptible to internal parasites and deworming should be considered for pastured or grazed sheep. Fecal testing is the best way to determine which parasites are present and can help you construct a deworming program. It is not advised to deworm your flock regularly as the parasites may become drug-resistant if dewormers are overused.

If you pursue organic certification, you will not be allowed to use most or all of the vaccines available. Check with your organic certification representative prior to using any vaccines.

CLOSED FLOCK

Once you establish your flock, the best way to maintain its health is to keep a closed flock. By not purchasing any outside animals you will keep your disease risk to a minimum. Your replacement females can be selected from within your flock and new acquisitions can be limited to rams. You may be able to grow your own rams from within, but this continued practice may lead to unacceptable increases in inbreeding or linebreeding. Rams typically spread fewer diseases than ewes, although they can harbor pinkeye, soremouth, footrot, and other infections and should be carefully and thoroughly examined prior to being placed with the ewes.

One overlooked situation is the lending of sheep, particularly rams, to other farms for breeding purposes. If this is done, be sure the other flock has a health status that is comparable to yours. Lending a ram for breeding in another flock and bringing him back into your flock can bring in unwanted problems. If you do this, make certain that he is isolated for an appropriate time and have him checked by your veterinarian before placing him back in with your ewes.

You should also be careful about allowing others to bring ewes to your farm for breeding to your ram. Unless the ewes of that flock have a similar health status, you may be bringing in potential problems.

GOOD FARM MANAGEMENT

There are other concerns affecting the health of your flock that are not sheep-related in the strictest sense and can be controlled by good farm management practices. Keeping rodents, cats, and other wildlife that harbor infectious agents away from your flock is good farm management. Cats can carry the toxoplasmosis protozoan, which can cause abortions in sheep if they ingest infected grains or stored hay. Vaccinating all cats on the farm will help prevent this problem and will ensure a healthy, stable population of felines.

One of the leading causes of abortion in sheep is toxoplasmosis which can be transmitted from infected farm cats. To prevent ewes from becoming infected, you should have your cats vaccinated and kept away from stored hay and grain. Maintaining a healthy adult population of cats may be just as important as keeping healthy sheep.

Maintaining a clean farm, reducing the impact of rodents, and commonsense farming practices can be helpful in your biosecurity program. More information and assistance are available from your county agricultural extension office.

FARM PRACTICES

Animal rights and animal welfare are two terms sometimes used interchangeably; both refer to proper animal care. However, the two terms have different meanings and supporting philosophies and have become a concern to anyone raising livestock. Animal-rights campaigns occasionally reach front-page news and are a subject with which all farmers must take time to familiarize themselves.

For generations, farmers have known that it is in their best interests to provide humane care to their animals. They understand that the comfort and thriving nature of their animals is enhanced when provided with such unspoken rights as food, shelter, light, clean barns, and the ability to move about freely. This is often referred to as animal welfare.

This stewardship has been appropriated in the last few decades by groups whose agenda has included the elimination of animal agriculture; they believe that humans have no right to use animals for their own purposes. Considering that only 1.6 percent of the U.S. population lives on farms today, there is no question that agriculture is getting smaller and urbanization is growing. Animal rights have become a social issue rather than a scientific issue. As these views spread and gain traction, it is up to livestock producers to address these perceptions and defend their humane practices not just scientifically, but morally as well. It is about what the animals represent and the values associated with raising them.

Young farm people can be placed in an uncomfortable position by animal-rights advocates, particularly at shows or other public venues where urbanites come into contact with rural residents. Helping young 4-H or FFA members understand how to approach these concerns will provide an educational opportunity to engage in this subject.

Training kits from various farm organizations and livestock groups are available to help you and your family understand the issues and how to respond to those who do not. Raising sheep that provide food or wool for other people is noble work and one that has given great satisfaction to many farm families.

Some diseases can be introduced and spread by shearing. Caseous lymphadentis is an infectious, contagious disease that is the third-leading cause of carcass condemnation in cull ewes. Request that your shearers disinfect their equipment between flocks and between sheep. You will limit the risk of spreading the disease by shearing the youngest sheep first.

177

Immediately removing placentas and fetal tissues from lambing areas will help prevent the introduction and/or spread of diseases. A ewe should not be allowed to eat her placenta as this can spread scrapies. Composting is a good method for disposing of waste products from lambing. Do not leave carcasses or placentas for dogs or wild animals to eat as this can attract predators to your premises. *Jan and Coby Schilder*

COUNTRY LIVING

People who own a farm in the country and people who own a house in the country have two different perceptions of the same area, with a wide gulf in between. The purpose of your farm is to produce agriculture, and the purpose of a rural home is to enjoy all the benefits of living in the wide-open spaces without having to contend with many urban issues including traffic, close neighbors, and noise at night.

With two different agendas at work in the same area, it is understandable that conflicts and disagreements can arise as to which agenda takes precedence. Many states now give farmers a basic right to farm without the fear of lawsuits brought by offended neighbors. In these cases, an agricultural operation is presumed not to be a nuisance to the neighbors even when new neighbors move in. If the farm operations are conducted in a reasonable manner, the neighbors can't legally complain. Landowners, residents, and visitors must be prepared to accept the effects of agriculture and rural living as normal and understand that they are likely to encounter a number of practices that area farmers have been and likely will continue doing in their normal farming practices. Some states have developed a list of specific annoyances that are not considered a legal nuisance to neighbors including odor, noise, dust, and the use of pesticides or other chemicals. As a rural landowner, you should take time to acquaint yourself with local ordinances that may affect your farming practices.

IN YOUR BACKYARD

It is often impossible to stop people from purchasing land near or adjacent to your farm. You have little control over what their expectations are for moving into your neighborhood. You may feel they should be grateful for having the chance to live where you do, just as you have decided to venture into farming for possibly similar reasons. But that may not be the position they take, and they may feel it is their right to be there without the inconveniences agricultural production imposes upon them. You may find yourself

confronted by disenchanted neighbors over something minor, perceived or otherwise, but there are ways of disarming a confrontational situation before it gets out of hand. Taking a proactive approach before it reaches this stage may be your best defense.

Inviting new neighbors to visit your farm and see it up close may be one way to build bridges with them. A friendly atmosphere where you can discuss your farming practices and procedures, the goals of your business, the values you hold for the land, and the reasons for involving your family in the farm may help offset their concerns as they witness your family's total involvement.

Using a positive approach, providing an occasion for implementing your biosecurity measures in their presence, and allowing them to see your humane sheep-raising program may make them more appreciative of the area in which they have chosen to live.

One of the best management practices for biosecurity is to maintain a closed flock and not bring any outside animals to your farm. Because of the advent of artificial insemination, you can eliminate the need for bringing an outsize ram into your flock.

SHOWING SHEEP

Organizations across the country such as 4-H and FFA offer activities that provide opportunities for youth to experience the pleasures and challenges of showing sheep in organized competition. Showing sheep can help youth develop skills and teach lessons that can be useful throughout life.

Many opportunities exist for showing sheep whether it is through 4-H, FFA, or Open Class where adult breeders compete. Learning to feed, handle, and care for sheep are some of the many positive experiences young people can gain from participating in showing. *Jan and Coby Schilder*

A sheep project combines many disciplines of animal husbandry, including feeding, grooming, handling, training, and exhibiting the sheep at a local, county, or state fair or a breed-sponsored or sanctioned show. Depending on the quality of the animal or the interest level of the member, regional and national shows provide excellent learning opportunities on a broader scale. Classes for showing sheep can include market lamb, breeding ewe, breeding ram, dairy ewe, wool-producing ewe, and others.

The opportunity to work with an animal and be solely responsible for its well-being over an extended period of time can have many residual, often intangible benefits for the 4-H or FFA member. While prizes, premiums, and recognition are part of the benefits and experience of showing sheep, young men and women can learn a variety of skills not readily attained in other ways. These skills range from the daily care and handling of animals to time management and responsibility. Showing places them among other young people of their age interested in sheep, as well as the organizations, and allows them to become acquainted with other FFA and 4-H members from the county and state. The composite experience can develop skills such as decision making, discipline, and patience—life lessons that can prove useful in later careers.

PROJECT OPTIONS

There are two types of projects generally available to 4-H, FFA, and other youth organizations: ownership or managerial. An ownership project involves a member purchasing a prospect ewe or lamb and assuming all costs during the length of the project and receiving all benefits including awards, honors, and total receipts at the final sale of the animal. A managerial project allows the member to raise, care for, and exhibit an animal without having to own it. This generally involves members whose parents, neighbors, or friends own the animal and allow the member to feed, train, and raise it for a specified period of time. This arrangement provides the member with the opportunity to learn how to properly care for a lamb or ewe without the purchase cost. In this case, the member is generally responsible for all expenses during the length of the project including feed, veterinary care, and insurance. In return, the member receives all awards or financial premiums won at any show where their sheep are exhibited. When the animal is sold to market, the owner receives a percentage of the sale price unless other arrangements between the owner and member have been made. Many different types of arrangements may be possible between owners and project members, and local 4-H or county leaders and FFA advisors can usually help find the best option for the member.

PERSONAL GROWTH

Ethics are an important part of showing animals and help keep the competition fair for all exhibitors. Participating in shows allows the member to learn about the ethics of raising and showing animals in competition.

Rules and codes of ethics have been developed for showing sheep regarding practices that are allowed and those that are discouraged or not tolerated. Any human

Lamb sweater blankets are used to keep sheep clean, keep light pressure on muscles, and remove wrinkles on fleece before a show. Blankets do not damage the fleece and are available many different sizes and colors.

The showmanship class is one of the most exciting for both young exhibitors and the audience. The exhibitors handle their sheep without halters or canes. Showmanship allows the exhibitors to demonstrate their expertise with handling their sheep and the grooming, feeding, and care that has gone into their project.

A market class involves the judge handling each of the sheep in the class. The judge uses his or her knowledge and experience in sheep breeding to assess which animal in the class has the greatest muscling in the loin, rump, and shoulder areas and places the animals according to their live evaluation.

manipulation that alters the physical appearance of the animal, aside from the wool coat, is prohibited. The reasons for these rules should be obvious. The alteration of the animal's physical structure, whether it is the bone or muscle, is detrimental to the well-being and health of the animal and cannot be tolerated for the sake of winning a prize.

The showing of sheep will generally take the appearance of both the exhibitor and the sheep into consideration, especially if participating in a showmanship contest. While the regular classes emphasize the productive aspects of the animal, such as its meat-producing abilities, showmanship is a separate class in which the exhibitor's ability to show the animal to its best advantage is the primary consideration. This class also takes into account the exhibitor's knowledge of his or her animal and correct procedures to handle the animal in the presence of a judge, as well as the interaction between one exhibitor and another and between the exhibitor and the judge.

Learning the ethics of fairness and properly handling and raising animals is basic to the emotional and mental development of the member. Understanding the ethical choices available and learning to make them is one of the most valuable lessons young people can learn from a 4-H or FFA project. Guidelines for understanding these ethical practices are available from breed associations, industry organizations, county 4-H extension offices, and school FFA programs.

DEVELOPING LIFE SKILLS

Because 4-H and FFA members assume total control of their sheep project, they can develop recordkeeping and budgeting skills. Maintaining sound records, including all costs, expenses, income, and health of their sheep, is part of the record-keeping process to complete the project evaluation at the end of the year. FFA and 4-H programs provide training and assistance to teach youth how to keep accurate records and serve as a model for later careers. Showing can be fun, but the lessons young men and women can learn from raising sheep are beyond financial measurement.

Sheep provide a unique venue for training when compared to other livestock. Dairy and beef cattle are halter-broke and tied up during their stays at fairs and exhibitions. Pigs are not physically handled but are trained using canes or PVC pipes to provide nonverbal direction. Sheep are manually handled by the exhibitor to move them from their holding pens to the show ring and for show-ring exhibition. Sheep need training so they become acquainted and comfortable with their handler. It is easier if this training begins while they are still young, unless the member starts with a ewe. More time will need to be spent handling a ewe that may not have been worked with prior to the project.

Showing sheep takes discipline by the exhibitor and involves the daily dedication to animal care and scheduling the time to provide proper nutrition, water, and housing to make sure the ewe or lamb is comfortable and well fed. This ensures adequate growth so the lamb or ewe is comparable in size to others within its own age group or class when shown.

Many states offer youth programs that encourage their participation in educational and exhibition events. These help young members broaden their knowledge of sheep production above and beyond the show ring. *Todd Taylor*

LEADERSHIP SKILLS

Leadership skills can be developed from showing sheep through the 4-H or FFA projects. Clubs and chapters provide adult supervision to help members learn how to develop leadership skills by allowing them to direct discussions and show younger members how to handle their own projects. Older members can demonstrate showing techniques they have learned and explain methods of raising animals that support ethics training.

Learning leadership skills, discipline, and patience at a basic level serves as a platform from which the member can build his or her personal integrity. This can have significant advantages when he or she later enters the job market. Having an understanding of animals and the ability to handle them correctly and ethically can provide

members with many career choices not available to their contemporaries.

Careers in animal-related or scientific fields such as the veterinary sciences, research, animal product sales, genetics, agriculture journalism, or education can be started with a project as simple as raising and showing sheep. People with experience in handling animals, ethics training, recordkeeping, and leadership skills are always in demand by employers. More information about the 4-H program is available from county 4-H extension offices. Information about the FFA program is available from a local high school chapter or your state FFA office.

SOME CAREERS AVAILABLE IN ANIMAL AGRICULTURE

Agriculture Engineer	Nutritionist
Agriculture Journalist	Researcher
Extension Educator	Salesperson
Farm/Ranch Manager	Teacher
Livestock Buyer	Veterinarian
Meat Cutter	Vocational Agriculture Instructor

Showing is not limited to youth. Many adult breeders also participate in shows as a form of promotion for their breed or their farm's stock breeding program. Many adults derive as much pleasure from showing as the youth. *Jan and Coby Schilder*

EXIT STRATEGIES

Change happens in life and on farms. For anyone living on a farm, there comes a time when a sale occurs, whether a complete sale precipitating an exit from farming or a partial sale of the assets, such as your sheep or machinery, and retention of the land. It may be a sale to another family member or a combination of these instances. As with any other aspect of your business, planning ahead for a sale will generally pay greater dividends than if you let events dictate your course of action.

Life and farming situations ebb and flow through the years and change is one constant facing anyone who decides to farm. Weather conditions, finances, personal goals, and agendas all factor into decisions made on a daily or yearly basis.

The reality is that a sale is likely to occur in your lifetime. There is no disgrace in this situation because farms have been bought and sold for generations as people have entered and exited farming. That may be the cold reality, but there is little doubt that leaving a farm can be an emotional time for you and your family. The reasons for leaving may include health issues, finances, changes in a family situation, or a desire to pursue other interests.

In many cases it may seem easier to get into a business than to get out. If you plan to exit from your sheep business, there are some basic similarities—and differences—to any other business that has an exit strategy. A discussion with your financial advisor may provide answers to help you successfully exit farming.

DON'T FEEL GUILTY

You do not have to feel guilty that you are leaving the farm or having a sale. Families who leave their farm too often develop a sense of guilt because they consider it a failure on their part. Perhaps financial problems contributed to the sale, but the fact that they spent time and effort in trying to make their business succeed should not be viewed as a failure. Farming can be a challenge even in the best of times. Market forces have a way of blowing down the best-built house and those who can withstand such events are sometimes more lucky than good managers.

SALE OPTIONS

Farms are different from many other businesses in that you have living assets—animals—that must be sold. They are similar to other businesses because there are also nonliving assets to be sold and tax considerations after a sale. There are several ways to handle the sale of your sheep or machinery and each has advantages and disadvantages. A thorough understanding of each option can save you money and minimize surprises.

If you have few sheep to sell, your choice of exit may be simply to market them through a packing house if they are of appropriate weight. Or you may try to sell them at public auction or privately to another sheep grower. Exiting a sheep-raising program is a relatively simple procedure, but ending it does not necessarily mean an end to your farming enterprise.

Exiting from a sheep-raising program may not be as difficult as getting started. The options of disposing of your sheep may include the same ones you used when you bought them, such as through public or private sale.

The conditions for having your sale may determine what happens next on your farm. If you sold all the animals from your farm, then the land will still be part of your assets. Your facilities can stay empty if the sheep are gone, but this will mean you will have to pay taxes on unused buildings. The buildings can be rented to another party to provide income. A written contract for renting to another party is good business and will keep any potential problems to a minimum. You can also rent out your land and still retain ownership and live on the farm. Another option for your farm if you discontinue raising sheep is to convert your land into a vegetable produce farm or other cash crops.

POSTSCRIPT

It should be obvious from these examples that the sale of your sheep does not need to be the end of your farming life unless you choose it to be. There have been many farmers who enter a sheep-raising business, leave, come back in, and then leave again. Some try to avoid the peaks and valleys of marketing cycles and use an entry and exit policy. Another reason for selling and going back into sheep production may have to do with wanting time away from the farm or health and personal issues. Selling your sheep does not have to be a traumatic experience if you accept the idea that a sale will happen eventually either with or without you in attendance. Planning ahead for a sale can take a lot of the emotional stress out of the decision. It can leave you with a healthier state of mind and a sense of satisfaction that you have accomplished the goals you set at the time you entered the sheep business.

Your farm can be rented to other parties for a specified period of time. This may allow you to re-enter into a sheep raising program later if you choose.

HELP AND ADVICE

You may have questions from time to time about how to handle certain problems that arise on your farm. Help is available to find answers to your questions, solutions to your problems, and new perspectives to consider.

COUNTY AGRICULTURAL EXTENSION SERVICES

The Agricultural Extension Service, usually located at your county seat, can provide help with answers and solutions. Extension agents are specially trained and have access to their state's university systems and research

Help and assistance in your sheep program can come in many forms. You can access information from county agricultural extension offices, university research stations, and technical schools.

departments. They are able to glean a large amount of information from these sources and pass it along to you.

Because of their extensive contacts across the country and around the world, extension agents receive information on traditional practices, as well as the latest innovations to arrive on the scene as new approaches make front-page news in farm publications. Although you may have the same access to electronic data, their background and experience can help narrow your focus, and their services are free.

TECHNICAL SCHOOLS

Technical schools provide information, training, and assistance by offering a hands-on approach. Classes offered by technical schools can supplement and provide training that may not be available through county agricultural extension offices. Their classes may help you prepare other areas of your farming business not directly related to livestock, such as accounting or financial assessments.

GRAZING NETWORKS

Grazing networks are an excellent way to get to know other farmers in your area or region who are developing grazing programs. These networks are informal groups with the purpose of helping beginning grazers solve problems, explore alternative methods, provide social contacts, and generally have fun while learning about the way other grazers handle their pastures, problems, and successes. Because these networks are made up of like-minded people, ranging from experienced to novice, their attitudes reflect helping one another rather than forcing views upon new members.

Grazing networks typically have a mix of livestock farmers in their membership with dairy, beef, pig, goat, and sheep owners among the most prevalent. The main focus of the grazing network is the pasture walk that typically takes place during the main grazing season from April to October. Pasture walks are typically held on a different farm each month and begin with a brief introduction by the host farmer or extension agent, with a short discussion prior to going into a pasture to walk and observe the conditions the host farm is using. Questions are encouraged. Topics covered can include pasture

Grazing networks are a good way to learn how others approach their pasture grazing programs and meet like-minded producers. They can be educational and social events and help broaden your knowledge about different farm-related subjects.

Sheep breed associations can offer information about their respective breeds, production data, animals for sale, breed auctions, show events, registrations, and membership. Universities and agriculture technical schools typically offer shearing classes where sheep owners can learn the art of shearing sheep. Contact your county agricultural extension office for information about seminars that may be offered near you. *Department of Animal Science, University of Wisconsin-Madison*

composition or what grasses are planted and being used for forage, fencing layout, how that particular system works for the owner, watering system, lanes, and the overall grazing management being used. Some networks use a series of walks on the same farm during the year to follow the farm's progress and learn how the host farmer handles the changes over the grazing season.

UNIVERSITY SYSTEMS

All land-grant colleges in the United States provide agricultural classes that are available to the public. The university system can provide research data on a variety of subjects. These reports may help in making decisions related to your farming practices.

STATE DEPARTMENTS OF AGRICULTURE

The United States Department of Agriculture (USDA) is the cabinet-level agency that oversees the vast national agricultural sector. Its duties range from research to food safety to land stewardship. Every state in the country has a department of agriculture that administers the programs of that particular state and operates under its statutes. These agencies have a wide range of booklets, pamphlets, and other publications available to help you understand the rules and laws, obtain licenses, and comply with regulations pertaining to your business.

UNLIMITED RESOURCES

The resources available to you are limited only by the amount of time you dedicate to researching them. The Internet has made huge quantities of information available that may apply to your situation. As with all things related to electronic data, it is best to check the authorities of the articles you use and to use care in providing information about your farm to outside requests.

Research conducted by universities or research stations aid sheep producers in many ways. Universities can offer extensive public assistance with access to numerous publications, research data, and personnel with expertise in many areas.

RESOURCES

Breed and Association Addresses

American Border Leicester Association
Di Waibel, Secretary/Treasurer
P.O. Box 947
Canby, OR 97013
(503) 266-7156
momfarm@canby.com
www.ablasheep.com

American Cheviot Sheep Society
Jo Bernard, Secretary/Treasurer
Rt. 1, Box 367
New Richland, MN 56072
(507) 465-8474
mrscheviot@myclearwave.net
www.cheviots.org

American Corriedale Association, Inc.
Marcia Craig, Executive Secretary
P.O. Box 391
Clay City, IL 62824
(618) 676-1046
info@americancorriedale.com
www.americancorriedale.com

American Delaine & Merino Record Association
Connie M. King, Secretary/Treasurer
59419 Walters Road
Jacobsburg, OH 43933
(740) 686-2172
kingmerino@alltel.net
www.admra.org

American Dorper Sheep Breeders Society, Inc.
Douglas P. Gillespe, Executive Secretary
P.O. Box 796
Columbia, MO 65205
(573) 442-8257
dorpers@ymail.com
www.dorperamerica.org

American Hampshire Sheep Association
15603 173rd Avenue
Milo, IA 50166
(641) 942-6402
info@hampshires.com
www.hampshires.com

American Karakul Sheep Registry
11500 Highway 5
Boonville, MO 65233
(660) 838-6340
aksr@iland.net
www.karakulsheep.com

American Polypay Sheep Association
Karey Claghorn, Secretary
15603 173rd Avenue
Milo, IA 50166
(641) 942-6402
info@polypay.org
www.countrylovin.com/polypay

American Rambouillet Breeders Association
1610 South State Road
Levelland, TX 79336
(806) 894-3081
contact@rambouilletsheep.org
www.rambouilletsheep.org

American Shropshire Registry Association
Becky Peterson, Secretary
41 Bell Road
Leyden, MA 01337
(413) 624-9652
shropsec@hotmail.com
www.shropshires.org

American Southdown Breeders' Association
Gary Jennings, Secretary
100 Cornerstone Road
Fredonia, TX 76842
(325) 429-6226
southdown@ctesc.net
www.southdownsheep.org

Barbados Blackbelly Sheep Association International
Mary Swindell
815 Bell Hill Road
Cobden, IL 62920
(618) 893-4568 or (618) 967-5046
secretary@blackbelly.org
www.blackbellysheep.org

Columbia Sheep Breeders' Association of America
Doug Gehring, Executive Secretary
1371 Dozier Station Road
Columbia, MO 65202
(573) 886-9419
dougkg@centurytel.net
http://columbiasheep.org

Continental Dorset Club
Debra Hopkins, Executive Secretary
P.O. Box 506
North Scituate, RI 02857
(401) 647-4676
cdcdorset@cox.net
www.dorsets.com

Cotswold Breeders Association
Tony Kaminski, Registrar
P.O. Box 441
Manchester, MD 21102
(410) 374-4383
kaminskicotswolds@mris.com
www.cotswoldsheepbreeders.com

East Friesian Members Dairy Sheep Association of North America
Larry Curtis, Secretary
HC 69, Box 149
Anselmo, NE 68813
(308) 749-2349
ewemilk@nebnet.net
www.dsana.org

Finnsheep Breeders Association
Cynthia Smith, Secretary
HC 65, Box 517
Hominy, OK 74035
(918) 885-1284
cindyusmith@yahoo.com

Jacob Sheep Breeders Association
Mickey Ramierz, Membership Secretary
2540 West Mulberry Street
Fort Collins, CO 80521
info@jsba.org
www.jsba.org

Katahdin Hair Sheep International
James Morgan
P.O. Box 778
Fayetteville, AR 72702
(479) 444-8441
khsint@earthlink.net
www.khsi.org

National Lincoln Sheep Breeders Association
15603 173rd Avenue
Milo, IA 50166
(641) 942-6402
kclaghorn@earthlink.net
www.lincolnsheep.org

National Tunis Sheep Registry, Inc.
15603 173rd Avenue
Milo, IA 50166
(641) 942-6402
kclaghorn@earthlink.net
www.tunissheep.org

Navajo-Churro Sheep Association
Connie Taylor, Secretary
P.O. Box 135
Hoehne, CO 81046
churrosheep@mac.com
www.navajo-churrosheep.com

North American Romanov
Sheep Association
Don Kirts, Secretary
P.O. Box 1126
Patashala, OH 43062
(740) 927-3098
narsa@columbus.rr.com
http://home.columbus.rr.narsa

North American Shetland
Sheepbreeders Association
15603 173rd Avenue
Milo, IA 50166
(641) 942-6402
secretary@shetland-sheep.org
www.shetland-sheep.org

North American Texel Sheep Association
Gayle Smith, Secretary
Stillmeadow Farm
740 Lower Myrick Road
Laurel, MS 39440
(601) 426-2264
stillmeadow@c-gate.net

North American Wensleydale Sheep
Association
4589 Fruitland Road
Loma Rica, CA 95901
(530) 743-5262
info@wensleydalesheep.org
www.wensleydalesheep.org

Scottish Blackface Sheep Breeders
Association
R. J. Howard, Secretary
1699 HH Highway
Willow Springs, MO 65793
(417) 962-5466

United Suffolk Sheep Association
P.O. Box 256
Newton, UT 84327
(435) 563-6105
UnitedSuffolk@comcast.net
www.u-s-s-a.org

U.S. Targhee Sheep Association
950 County Line Road
Fort Shaw, MT 59443
(406) 467-2462
roeder@3rivers.net
www.ustargheesheep.org

Other Organizations
American Livestock Breeds Conservancy
P.O. Box 477
Pittsboro, NC 27312
(919) 542-5704
www.albc-usa.org

FFA
6060 FFA Drive
Indianapolis, IN 46282
(317) 802-6060
www.ffa.org

4-H
1400 Independence Ave. S. W. Stop 2225
Washington, D.C. 20250
(202) 720-2908
www.4husa.org

GLOSSARY

• •

Abortion: Premature loss of a pregnancy.
Accelerated lambing: When a ewe lambs more than once a year.
Afterbirth: Placenta and fetal membranes expelled from the uterus after the lamb(s) are born.
Antibody: Substance that helps fight disease. Colostrum is high in antibodies.
Apparel wools: Wools manufactured for use as clothing.
Artificial insemination (AI): Process where semen is placed within a female's uterus by artificial means rather than through sexual intercourse.
Average daily gain: Amount of weight that a lamb gains each day.

Balanced ration: Food for animals that includes all the daily required nutrients.
Banding: Process of applying rubber bands to the tail or scrotum for docking or castrating.
Bloat: Digestive disorder of ruminants, usually characterized by an excessive accumulation of gas in the rumen.
Bottle jaw: Edema or fluid accumulation under the jaw; a sign of parasite infection.
Breech birth: Birth when the lamb is presented backward with the rear legs tucked under.
Breed: Group of sheep with similar characteristics and the same ancestors.

Breeder: Owner of the lamb's dam at the time she was bred.
Breeding season: Season of sexual activity when ewes are bred.
Breeding soundness examination: Physical examination of a ram used to determine if he is capable of impregnating ewes.
Broken mouth: Condition where a sheep has lost some of its teeth.
Byproduct: Made from parts leftover from major processes.

Carcass yield: Carcass weight as a percentage of live weight.
Carding: Process of moving wool fibers between two surfaces covered with wire pins to detangle and align the fibers in preparation for spinning.
Castration: Removing the testicles from male animals.
Clean fleece weight: Weight of fleece after it has been washed and scoured.
Clean price: Price paid per pound of clean wool.
Colostrum: The first milk produced by the mother after it gives birth; high in antibodies.
Combing: Straightening fibers into parallel strands using combs.
Commercial flock: Farming enterprise that sells pounds of lamb and wool.
Composting: Process of organic wastes decomposing naturally.
Concentrate: Feed that is high in energy, low in fiber content, and highly digestible.
Conformation: Physical shape and design of the animal.

Continuous grazing: Leaving sheep on a grazing area through the entire time grazing is available.

Creep feeding: Giving lambs extra feed when they are still nursing.

Crimp: Natural waviness of a piece of wool.

Crossbred: Animal whose parents are of two different breeds.

Crotching: Removing the wool from around the tail and between the rear legs of a sheep.

Cryptorchid: Male lamb with testicle(s) retained in the abdominal cavity.

Cud: Regurgitated food.

Culling: Process of removing animals that are below average in production or unsound.

Dags: Wool contaminated with feces that adhere to or have been clipped from the rear end of the sheep.

Dam: Female parent.

Docking: Removing long tails from baby lambs.

Drench: Method of giving liquid medicine.

Dressing percentage: Percentage of the live animal that ends up as carcass.

Dual purpose breeds: Breeds used for more than one purpose; such as meat and milk.

Dystocia: Difficulty in giving birth or being born.

Estrus: Period of time when a female is sexually receptive to a male; also referred to as heat.

Estrus cycle: Time from one estrus to the next; about 17 days in ewes.

Ewe: Female sheep.

Ewe breeds: Usually white-faced breeds of fine-wool type.

Ewe lamb: Female sheep less than one year old and is usually not bred.

Expected progeny difference: Estimate of how an individual's offspring will perform compared to the average of the flock.

Fabrication: Process of cutting lamb carcasses into wholesale cuts.

Fatten: To feed for slaughter, make fleshy, or plump.

Feed: Food given to sheep so they have all the essential nutrients.

Feeder lamb: Sheep under one year of age that is not ready for market but will make good gains if placed on feed.

Feedlot: Area where animals are fed or finished for market.

Feedstuff: Material used for feed such as silage, corn, or other plant products.

Fever: Body temperature above normal.

Fiber: Single piece of wool.

Fitting: Clipping, washing, and grooming of animals for show.

Fleece: Wool from one sheep.

Fleece weight: Weight of shorn wool from one animal.

Flock: Small group of sheep.

Flushing: Practice of feeding and managing the ewes so they are gaining weight when the breeding season begins.

Footrot: Contagious disease of sheep's feet caused by specific bacteria.

Gestation: Pregnancy; fetus development period between fertilization and birth.

Grade: Any animal, not purebred, that has the major characteristics of a breed.

Graft: Transfer a lamb to a ewe that is not its mother.

Grain: Feeds like corn, wheat, and barley that are high in energy.

Grease fleece weight: Weight of fleece after it has been shorn and prior to washing and scouring.

Grease price: Price paid per pound of grease wool.

Grease wool: Shorn wool that has not been washed.

Gummer: Sheep that has lost its teeth; a show of age.

Heat: See estrus.

Heritability: Likelihood of certain traits being passed genetically to future offspring.

Heterosis: Increased performance from crossbred individuals relative to the average performance of the parent purebreds.

Hybrid vigor: Increase in the performance resulting from crossbreeding.

Inbreeding: Mating of individuals more closely related than the average of the breed.

Jug: Pen where a ewe and her newborn lambs are put to bond.

Lamb: Sheep under one year of age; meat from young sheep; or to give birth.

Lanolin: Wool grease.

Lethal defect: Death of an animal caused by its genes.

Linebreeding: Use of closely related relatives in a breeding program.

Market weight: Weight of the animal when sold for processing.

Mastitis: Inflammation of the mammary glands.

Mixed grazing: Grazing by two or more species on the same land.

Mouthing: Determining the age of sheep by examining the teeth.

Mutton: Meat from an older sheep.

Muzzle: Nose of a sheep.

Offspring: Animals born to a parent.

Open: When a ewe is not pregnant.

Outcrossing: Mating of individuals less closely related than the breed average.

Out-of-season breeding: Practice of breeding ewes in spring or other time of year not typical of your major breeding times.

Paddock: Enclosed area for grazing animals.

Parturition: Act of giving birth.

Pastern: Region of the foot or leg between the hoof and dewclaw.

Pasture: Grasses or legumes grown in fields for grazing animals.

Pinkeye: Condition where the membranes lining the eyelids and eye covering become infected.

Poll: Head of the sheep.

Polled: Animal that naturally has no horns.

Price spread: Difference between the farm and retail prices.

Progeny: Offspring of the ram or ewe.

Prolificacy: Reproductive performance measured by number of lambs born per ewe lambing.

Purebred: Animal descended from a line of ancestors of the same breed; may or may not be registered.

Quarantine: Separation of a diseased or exposed animal from other animals in order to prevent the spread of a contagious disease.

Ram: Intact male sheep.

Ration: Amount of feed eaten or provided within a 24-hour period.

Registered: Purebred animals whose pedigrees are recorded in the breed registry.

Rotational grazing: Program where animals are moved from one grazing area to another.

Scouring: Process where grease (lanolin) and dirt are removed from wool.

Scrapie: Fatal brain disease in sheep.

Scurs: Small, rounded portions of horn tissue attached to the skin at the horn pits of polled animals.

Settled: Indications that an animal has become pregnant.

Shear: Act of cutting wool off the sheep.

Short day breeder: Animal that begins its breeding season as the days get shorter.

Showmanship: Presenting an animal at a showing including proper fitting of the animal, showing, and exhibitor appearance.

Silage: Crop that has been turned into animal feed through fermentation.

Single: Only one lamb born to a ewe.

Sire: Male parent.

Skirting: Removing the stained, unusable, or undesirable portions of a fleece.

Spin: To work natural fibers into thread or yarn.

Staple length: Length of the wool fiber in the fleece.

Stillborn: Lamb that is not alive at birth.

Stocking density: Relationship between the number of animals and area of land available.

Stocking rate: Number of animals grazing a unit of land for a specified period of time.

Tagging: Practice of shearing wool from udder and dock area of the ewe.

Total ewe productivity: Pounds of lamb weaned per pound of ewe.

Triplets: Three lambs born to a ewe at one time.

Twins: Two lambs born to a ewe at the same time.

Udder: Mammary glands, including the teats.

Vaccine: Injection given to animals to prevent or cure diseases.

Weaning: Removing a lamb from the ewe to stop nursing.

Wether: Male sheep that has been castrated.

Wool pool: Collection point for producers to sell their wool.

Wool yield: Pounds of clean wool as a percentage of pounds of grease wool.

Woolen: Yarn made from fibers that are one to three inches long and that have been carded.

Worsted: Wool yarn of long staple with fibers that have been combed prior to spinning; combing produces more parallel fibers than carding.

Yearling: Animal between one and two years of age.

Yolk: Natural grease covering on the wool fibers of the unscoured wool.

INDEX

· ·